Solar Photo
Guide for tl
Exam

This book explains the science of photovoltaics (PV) in a way that most people can understand using the curriculum which reflects the core modules of the NABCEP Entry Level Exam. Whether or not you are taking the NABCEP Entry Level Exam, learning the material covered in this book is the best investment you can make towards your place in the solar industry.

Providing complete coverage of the NABCEP syllabus in easily accessible chapters, the book addresses all of the core objectives that will aid in passing the PV Entry Level Exam, including the ten main skill sets:

- PV Markets and Applications
- Safety Basics
- Electricity Basics
- Solar Energy Fundamentals
- PV Module Fundamentals
- System Components
- PV System Sizing Principles
- PV System Electrical Design
- PV System Mechanical Design
- Performance Analysis, Maintenance and Troubleshooting.

You will learn how to survey a site, how to use the tools that determine shading and annual production, and the importance of safety on site. The text includes technical math and equations that are suitable and understandable to those without engineering degrees, and are necessary in understanding the principles of solar PV.

This study guide is written by **Sean White**, an IREC-Certified Solar PV Master Trainer, Electrician, NABCEP-Certified PV Installation Professional, Professor and Installer. Sean has prepared thousands of students to take the NABCEP Solar PV Entry Level Exam.

Solar Photovoltaic Basics: A Study Guide for the NABCEP Entry Level Exam

Sean White

Routledge
Taylor & Francis Group

LONDON AND NEW YORK

First published 2015
by Routledge
2 Park Square, Milton Park, Abingdon, Oxon OX14 4RN

and by Routledge
711 Third Avenue, New York, NY 10017

Routledge is an imprint of the Taylor & Francis Group, an informa business

British Library Cataloguing-in-Publication Data
A catalogue record for this book is available from the British Library

Library of Congress Cataloging-in-Publication Data
White, Sean (Electrical engineer)
Solar photovoltaic basics : a study guide for the NABCEP entry level exam / Sean White.
pages cm
1. Photovoltaic power systems--Examinations--Study guides. I. Title.
TK1087.W45 2015
621.31'244076--dc23
2014011466

ISBN: 978-0-415-71335-1 (pbk)
ISBN: 978-1-315-77011-6 (ebk)

Typeset in 11.5/11.5pt Rotis by
Servis Filmsetting Ltd, Stockport, Cheshire

Printed and bound in Great Britain by
TJ International Ltd, Padstow, Cornwall

Contents

Preface

Photovoltaics (PV) is a method of making electricity from light with a semi-conductor, usually silicon. Silicon is the same material from which a computer chip is made.

The solar industry is growing at light speed and now is time to get involved, to take part in reducing pollution and to take advantage of a great career opportunity.

By passing the NABCEP Entry Level Exam, you will have demonstrated that you are solar PV smart and ready to work in the fastest growing industry in the world.

This book is intended to help students learn the material on the exam in a reasonable time, while not getting overloaded with too much information. We will save the advanced material for another book, so you do not get overwhelmed.

Some of the more difficult material, such as voltage temperature calculations, will be covered multiple times in different ways throughout the book. This book is intended to be read and retained. Reading this book will be the most efficient time spent by busy people preparing to pass the NABCEP Entry Level Exam.

The NABCEP PV Entry Level Exam covers material, which is organized at www.nabcep.org into ten learning objectives:

1. PV Markets and Applications
2. Safety Basics
3. Electricity Basics
4. Solar Energy Fundamentals
5. PV Module Fundamentals
6. System Components

7.   PV System Sizing Principles
8.   PV System Electrical Design
9.   PV System Mechanical Design
10.   Performance Analysis, Maintenance and Troubleshooting.

The chapters in this book will match the ten learning objectives.

There will be a special section at the end of the book focusing on voltage and temperature calculations, practice exam questions and definitions.

PV markets and applications

KEY CONTRIBUTIONS TO THE DEVELOPMENT OF PV TECHNOLOGY

1839: Edmond Becquerel discovered the photovoltaic effect.
1905: Einstein described the photoelectric effect and how light (photons) can excite electrons.
1922: Einstein received Nobel Prize for describing photoelectric effect.
1954: Bell Labs developed the "Bell Solar Battery".

> The "Bell Solar Battery" is what is technically called a solar module today.
>
> Often people incorrectly call a solar module a solar panel.

1958: First solar powered satellite sent into space by US Navy. The Vanguard 1 is currently the oldest man made object in space.
1999: World total installed PV capacity 1 GW
    1 Gigawatt = 1000 MW (megawatts)
    1 MW = 1000 kW (kilowatts)
    1 kW = 1000 W (watts)
2012: World total installed PV capacity 100 GW, 31 GW of which was installed in 2012.

Figure 1.1 1956 PV advertisement from Bell Labs

Something New Under the Sun. It's the Bell Solar Battery, made of thin discs of specially treated silicon, an ingredient of common sand. It converts the sun's rays directly into usable amounts of electricity. Simple and trouble-free. (The storage batteries beside the solar battery store up its electricity for night use.)

Bell System Solar Battery Converts Sun's Rays into Electricity!

Bell Telephone Laboratories invention has great possibilities for telephone service and for all mankind

Ever since Archimedes, men have been searching for the secret of the sun.

For it is known that the same kindly rays that help the flowers and the grains and the fruits to grow also send us almost limitless power. It is nearly as much every three days as in all known reserves of coal, oil and uranium.

If this energy could be put to use – there would be enough to turn every wheel and light every lamp that mankind would ever need.

The dream of ages has been brought closer by the Bell System Solar Battery. It was invented at the Bell Telephone Laboratories after

long research and first announced in 1954. Since then its efficiency has been doubled and its usefulness extended.

There's still much to be done before the battery's possibilities in telephony and for other uses are fully developed. But a good and pioneering start has been made.

The progress so far is like the opening of a door through which we can glimpse exciting new things for the future. Great benefits for telephone users and for all mankind may come from this forward step in putting the energy of the sun to practical use.

BELL TELEPHONE SYSTEM

TYPES OF PV SYSTEMS AND THE BASICS OF HOW THEY OPERATE

GRID-TIED, AKA UTILITY-INTERACTIVE PV SYSTEMS

Grid-tied PV systems are connected directly to and synchronize with the utility. They are the most popular type of system. The **inverter is sized based upon the size of the PV array**.

The main components of a utility-interactive PV system are:

- Solar modules
- Inverter (no batteries or chargers).

Grid-tied PV systems have to be able to disconnect from the grid whenever the grid is down or not within specifications. This is called **anti-islanding** and means that the inverter cannot operate alone as an island of power. If a grid-tied system did feed the grid when the grid was down, it could be dangerous to utility workers who are fixing the problem. Some solar customers are surprised to find out that their utility-interactive PV systems will not work during a power outage.

OFF-GRID, AKA STAND-ALONE, AKA BATTERY-BASED PV SYSTEMS

PV systems that work independent of the utility grid. Usually used for remote homes. Stand-alone PV systems are designed to fulfill all of the electricity requirements.
There are two basic types of stand-alone PV systems.

> Direct current is the electricity from a solar module or a battery with a positive and a negative connection. Alternating current is what comes from your house and alternates very fast between positive and negative.

AC coupled systems   (ac = Alternating Current)
DC coupled systems   (dc = Direct Current)

DC coupled systems are the most common and simple off-grid systems. Their main components are:

- Solar modules
- Charge controller (prevents battery over- and under-charging)
- Battery
- Inverter.

AC coupled systems are complicated and less common. They use two types of inverters: **battery inverters** to create voltage so that **grid-tied inverters** can work when there is no utility. Their main components are:

Figure 1.2 DC coupled PV system

Stand Alone/Battery/Off–Grid

PV

Charge
Controller

Batteries

Inverter

DC coupled
system

AC loads

DC loads

- Solar modules
- Grid-tied inverters
- Off-grid inverter/charger
- Batteries.

HYBRID PV SYSTEMS

Hybrid PV systems include another source of power.
Typical other sources of power include:

- Generator (internal combustion engine)
- Wind turbine
- Micro-hydro (small hydroelectric turbine),

Figure 1.3 SMA Sunny Island AC coupled PV system

Loads

Sunny WebBox

Sunny Boy PV x

Sunny Boy PV 1

Windy boy

DC-Loads

Charge controller

PV

DC/DC converter

Fuel Cell

Battery current sensor

Battery

AC1 (Loads/Sunny Boys)

Com

Relay1/2

BatVtgOut

DC

BatCur

DigIn

AC2 (Gen/Grid)

Relay1/2

BatTmp

Battery temperature sensor

Utility

Generator

Description:
- - - Control voltage
*) DC supply contactor
   (not included in delivery)

GRID-TIED-BATTERY-BACKUP, AKA MULTIMODAL OR BIMODAL PV SYSTEMS

Grid-Tied-Battery-Backup (GTBB) PV systems can **operate as grid-tied systems and off-grid** systems. They are typically the most complex systems to design.

The inverters will produce as much power as possible when operating in utility-interactive mode. When the utility is interrupted, the inverters disconnect from the grid and switch to stand-alone mode and make as much power as the loads require. (**Loads are devices that consume electricity**.)

Multimodal systems have to disconnect from the grid completely when the grid is down, but still have to feed power to the house. These systems usually power a subpanel of specific loads and not the entire house.

DIRECT, AKA DIRECT-COUPLED PV SYSTEM

This is the **simplest type of PV system**. The **only components are PV and a load** (usually an electric motor).

A good example of a direct-coupled system is a solar attic fan. A solar attic fan consists of a PV module and a fan. When the sun is out, the fan works, when it is brighter, the fan works better, which is convenient, since we need a fan more when the sun is out.

Another common direct-coupled system is water pumping. In sunny times, more water is needed and water can be stored with elevation and used at night.

> Direct Water Pumping System Variation:
>
> Often times, water pumping systems use a linear current booster (LCB) which increases current and sacrifices voltage at times of low light, such as mornings.

There are no direct-coupled lighting systems, since when the sun is out, the direct sunlight is the most efficient light.

SELF-REGULATING PV SYSTEM

A self-regulating system is a stand-alone PV system **without a charge controller**. In most cases, not having a charge controller would damage a battery by under- and over-charging the battery.

According to the National Electric Code, a self-regulating PV system has to be designed so that it will **not charge over 3 percent battery capacity in an hour**. This way the battery will not be over-charged. Over-charging a battery can not only damage the battery, but also can split water molecules into hydrogen and oxygen, which is an explosive combination.

There are no safety issues with under-charging batteries, but as anyone with a car knows, letting a battery die is not good for the life of the battery, so self-regulating systems are designed with loads that are small relative to the size of the battery and PV, so that they can survive dark winter days.

A good example of a self-regulating PV system is a coastguard buoy, which has a big battery bank relative to a modest blinking light.

ADVANTAGES AND DISADVANTAGES OF PV

PV works when the sun is up and that is when people use electricity the most. We say that it generates at "peak" times, which makes it most valuable. Other conventional forms of energy production can be easily stored, however when we are burning things to make electricity, we are doing something that is bad for the environment.

Wind power is good when mixed with solar, because wind can work at night, however most people would rather live in sunny places than windy places. Also wind is intermittent, and sunshine is much more predictable. Even on cloudy days, solar systems make power. Solar works 365 days per year.

One of the problems that we have with solar is storage of power so that we can use the solar energy at night. As solar is adopted early and since solar energy is less than 1 percent of the energy produced on the grid, storage is not such a big deal; however as the industry grows exponentially, storage becomes more important. Energy storage is becoming more important in places where solar PV is more saturated, such as in Germany and Hawaii. As the need grows, production increases and the price drops with the development of new technologies. The price of PV has dropped ten fold in the last ten years.

When electricity is made where people live, we call that distributed generation (DG) and DG means that we do not have·to transmit electricity over large distances. When electricity is transmitted, there are losses. Also, with electricity produced nearby, we do not have to pay for land and infrastructure such as substations, high voltage power lines and transformers.

Utility scale solar farms are large solar projects; often in the desert where power is transmitted over long distances, such as is the case for most conventional power generation. This will add costs for substations and transformers, which often add millions of dollars to the project costs.

Most of the electricity in the world is made from coal. Heat from burning coal creates steam and that steam will spin equipment and make power. This is similar to the technology of a steam locomotive. Burning coal releases mercury and greenhouse gases into the atmosphere.

Nuclear power plants also use the heat of splitting radioactive material to make steam. As we know in the cases of Fukushima and Chernobyl, when we cannot keep the radioactive uranium cool, we can have a meltdown.

Burning natural gas, which is becoming popular due to the increased amounts of gas found using hydraulic fracturing (fracking), causes greenhouse gases to be released into the atmosphere. Also, the gases from leaked natural gas are much worse at trapping heat than CO_2. Additionally, fracking pollutes the domestic water supply, since chemicals are injected deep into the earth.

Solar power is a safe form of nuclear energy. We are using fusion reactions that are 93 million miles away to make light that we then convert to electricity with photovoltaic modules.

Much of the energy that we produce in the world is wasted and conserving energy is often a better value proposition than solar PV. One of the best energy investments one can make is on an efficient light bulb. Better yet, natural sunlight makes light and a properly designed building can take advantage of sunlight for lighting, heating and cooling. Using architecture for solar energy is called "passive solar".

PV on a rooftop can block sunlight from hitting the roof and keep a building cooler, lowering an air conditioning bill, even if the system is not turned on.

BENEFITS OF DIFFERENT TYPES OF PV SYSTEMS

Rooftop systems put the means of energy production on the roof, where energy is used, and protect the building from UV rays. Sometimes the building materials can be made from PV and this is called Building Integrated PV or BIPV. Since **BIPV is part of the building**, as in the case with a solar roof tile, then it will not have the same airflow underneath it as a regular PV system. When a PV system is hotter like this, the PV does not work quite as well as if it were operating cooler. **BIPV is not as efficient as regular PV because it is hotter** and it is more expensive, since regular solar modules are mass-produced to bring the cost down.

Solar grid-parity is when energy produced by PV is less expensive than electricity produced by conventional means. Grid parity is happening in different places in the world where there is a combination of factors, such as expensive conventional electricity, good sunlight (solar resource) and low installation and permitting costs. Hawaii has expensive conventional energy and good sunlight. Germany has low installation and permitting costs. As more places reach grid-parity, mass production brings the price of PV systems down more and the grid-parity map grows!

Ground mounted PV systems are easy to install, but they take up valuable real estate or even compete with endangered species.

Figure 1.4 97MW solar farm, Sarnia Ontario Canada
Source: Photo © Sean White 2010. At the time it was the world's largest PV system.
System consists of approximately 1.3 million First Solar thin film PV modules.

Safety basics

Chapter 2

The two most common safety questions that you may run into on any construction-related exam have to do with fall protection. This is what you need to know:

1. **6 feet** is the height at which you need fall protection.
2. **1:4 ladder ratio**.

Figure 2.1 1:4 Ladder ratio

Top of ladder at least 3' above contact with roof

1:4 ladder ratio

4

1

MORE SAFETY

Common sources for safety information are:

1. **OSHA** (Occupational Safety and Health Administration). There are federal and state OSHA requirements. State requirements cannot be more lax than federal.
2. **National Electric Code (NEC),** which is published by the National Fire Protection Association (NFPA). The NEC is what electricians use to make sure their systems are safe. It is also what the **Authority Having Jurisdiction** (AHJ) will use as the rules to follow when inspecting electrical systems.

> The AHJ is typically the city or county building department or inspector for most PV projects, but also includes anyone who has jurisdiction over a project and can include the utility or the state. The AHJ will interpret the NEC and adopt a version of the NEC.

> The NEC will tell us which wire to use. If a lot of current flows through a wire, it will heat up. If it gets too hot, it will melt the insulation around the wire. If the wire is in a hot place, it can take less current than the same wire in colder conditions. The **ability of a wire to carry current is called ampacity**. The hard NABCEP PV Installation Professional Exam is an open book NEC exam. The exam will require test takers to use tables and difficult to understand code to demonstrate a working knowledge of wire sizing and PV design.

More ladder safety:

- Rungs (steps) of a ladder are about one foot apart.
- The ladder should be 3 steps (3 feet) above the top of the roof or object you are leaning ladder against.
- Ladder should be secured to the roof at the top.
- Painting a wooden ladder can be unsafe and hide ladder defects.

- Metal ladders conduct electricity so electrical workers should use heavier fiberglass ladders.
- Use an extension ladder to climb on a roof, not an A-frame ladder.
- Keep 3 points of contact on a ladder. Carrying objects up a ladder in one hand can be dangerous!

ARCING AND ARC FLASHES

Electrical **voltage is the hydraulic (water) analogy of pressure** and when voltage is higher, so is the chance of a spark or an arc.

In a PV system, if there is a small gap in a circuit, then we can have an arc where electrons will travel through the air. Many newer inverters are equipped with arc fault detection. If there is an arc, then the inverter should shut off. An arc is a "plasma discharge" and is a very hot fire hazard.

An arc flash is an explosion when a great amount of energy is released with an arc. When working with larger systems and close to the potential spark/arc, you need to have proper arc flash personal protective equipment, which is like a space suit. Arc flashes are hotter than the sun. Often the more dangerous source of power is the utility rather than the PV system. The utility has more potential than our PV systems.

> **PPE is personal protective equipment** and can include arc flash suits, earplugs, goggles, aprons and personal fall arrest systems (PFAS). PFAS are usually the most important types of PPE protection gear in the solar industry.

OVERCURRENT PROTECTION DEVICES

Fuses and circuit breakers (overcurrent protection devices) are used to prevent fires. On the backside of a PV module, there is a label that says the **maximum series fuse rating**. The fuse will **open the circuit (turn off)** if there is too much current in order to protect wires and equipment.

Figure 2.2 Arc flash PPE
Source: Courtesy Honeywell

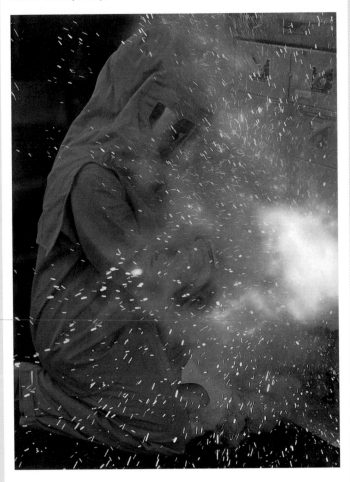

Figure 2.3 California Department of Public Health

No.24

CALIFORNIA FATALITY ASSESSMENT & CONTROL EVALUATION PROGRAM

CALIFORNIA DEPARTMENT OF PUBLIC HEALTH

SOLAR PANEL INSTALLER FALLS OFF ROOF!

A 30-year-old solar panel installer, Richard P* died after he fell 45 feet off the roof of a three-story apartment building. He was part of a three-man crew working to install solar panels on a sloped roof. Richard walked backward and stepped off the roof while checking the position of some brackets. No one was wearing personal fall protection equipment and there was no other fall protection system in place.

15

FALL PROTECTION

Falling is the number one cause of death in the construction industry.

Types of fall protection:

1. Guard rails
2. Safety nets
3. **Personal Fall Arrest System (PFAS)**.

The PFAS is the most common type of fall protection used in the residential solar industry.

PFAS consists of:

1. Harness
2. Shock absorbing lanyard
3. Lifeline
4. Anchor.

Figure 2.4 Harness
Source: Courtesy of Honeywell

Figure 2.5 Shock absorbing lanyard
Source: Courtesy of Honeywell

Figure 2.6 Anchor must hold 5000 lbs
Source: Courtesy of Honeywell

Figure 2.7 Keep the lifeline as short as possible
Source: Photo by Sean White at a very large solar factory near Shanghai China

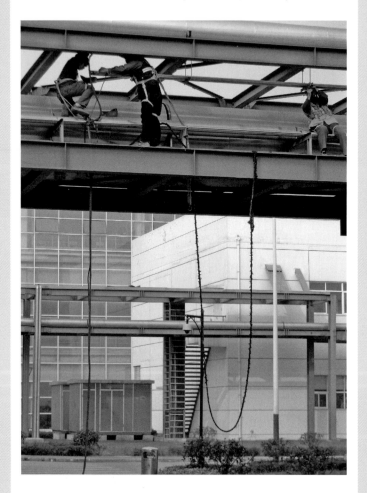

Figure 2.8 Fall protection must also be used around skylights. Skylights offer a false sense of security
Source: California Department of Public Health

DEADLY SKYLIGHTS!

SOLAR ENERGY AND WAREHOUSE WORKERS KILLED

A 46-year-old electrical worker, Luke J.*, and a 56-year-old warehouse worker, Roberto M.*, both died after falling through skylights while working on a roof. Luke was carrying a solar panel and tripped on the ledge of the skylight and fell backward through the skylight. Roberto was on the warehouse roof repairing a broken air conditioner. It is not known exactly how Roberto fell through the skylight.

Figure 2.9 Two broken First Solar PV modules next to skylight where solar installer died
Source: California Department of Public Health

The skylight that the worker fell through

There are some types of PPE that are more important than other types. Fall protection is the most important PPE in the solar industry. Goggles are more important than an apron when working with lead acid batteries. Your eyes are more important than your clothes.

LOCKOUT-TAGOUT (LOTO)

When you turn off your solar system and then go on the roof, you want to make sure that you can lock the system in the off position, so that nobody turns the switch on and electrocutes you.

Bodily safety

When you are working in the sun where there is no shade, protect yourself from the sun with sunscreen, drinking plenty of water and wearing a hat.

Also, with all safety precautions in place, people still get hurt and die, so above all, be extra careful!

Figure 2.10 Lockout-tagout kit
Source: Image provided Courtesy of Ideal Industries, Inc.

Electricity basics

In order to be competent with PV, we need to be competent with electricity.

Figure 3.1 Ohm's law wheel

$$I = \sqrt{\frac{P}{R}}$$

$$I = \frac{P}{E}$$

$$I = \frac{E}{R}$$

$$E = \sqrt{P \times R}$$

$$E = \frac{P}{I}$$

$$E = I \times R$$

Amps **I**

Volts **E**

Power **P**

Ohms(Ω) **R**

$$P = E \times I$$

$$P = I^2 \times R$$

$$P = \frac{E^2}{R}$$

$$R = \frac{E}{I}$$

$$R = \frac{E^2}{P}$$

$$R = \frac{P}{I^2}$$

POWER AND ENERGY

Laypeople and even newspaper reporters often use power and energy interchangeably. This is not correct and not the case for solar professionals.

Power is a rate, just like speed is a rate.

Power × Time = Energy
Speed × Time = Distance

When we leave the lights on, the cost depends upon how long we leave the lights on.

Power is measured in watts and the way to remember that is that the middle letter in po**w**er is a W for Watts.

Just like 1000 meters is a kilometer, 1000 watts is a kilowatt or kW.

The metric system as used in the solar industry.

1,000 = kilo = k
1,000,000 = mega = M
1,000,000,000 = giga = G

We are just moving the decimal 3 places.

We talk about watts when we often speak of a 250W solar module.

When we talk about the twenty 250W modules on our rooftop, we will call that

20 × 250W = 5000W
5000W/1000 watts per kW = **5kW**

When we talk about bigger commercial PV systems, we can talk of 4000 250W modules

4000 modules × 250 watts per module = 1,000,000 watts
1,000,000 watts/1,000,000 watts per MW = **1MW**

When converting, you are going to either multiply or divide by the conversion factor. It should be obvious if you multiplied instead of divided or vice versa. Make sure to use common sense rather than just memorization.

When we talk about the amount of solar in the United States, we will say that the US has over 13GW of PV operating as of 2013. That number usually doubles about every 18 months.

> Be sure to fully understand the following definitions and relationships. Go back a few times a day if you have to. Bookmark this page or fold the corner.

Simple definitions:

- **Voltage: Electrical pressure**
- **Current: Electrical flow**
- **Power: Rate at which electricity is used**
- **Energy: Amount of electricity used**

Relationships:

- **Power = Voltage × Current**
- **Energy = Power × Time**

Units:

- **Voltage is measured in Volts = V**
- **Current is measured in Amps = A**
- **Power is measured in Watts = W**
- **Energy is measured in Wh or more often kWh**

Symbols

- **Current = I (I for Intensity)**
- **Voltage = V or E**
- **Power = P**

How to do the math required to be entry level PV proficient:

$$W = V \times I$$

To solve for V, we put I under W

$$V = W/I$$

To solve for I, we put V under W

I = W/V

> Whatever symbol is alone in simple algebra will end up on top.
>
> If A × B = C then A = C/B

HOW TO CALCULATE ENERGY USAGE FOR AN OFF–GRID FOREST SERVICE FIRE LOOKOUT CABIN:

You have a 12V 2A light bulb that will run for 3 hours per day.

You also have a 100W radio repeater that will be on 100% of the time.

How much energy in kWh does the cabin use in a day?

Answer:

12V × 2A = 24W

24W × 3 hours = 72Wh of energy per day for the lights.

100W × 24 hours per day = 2400Wh repeater

Total energy = 2400Wh + 72Wh = 2472Wh/day

Figure 3.2 Simple algebra triangle. This triangle works with all simple equations. Cover up what you need to solve for. In this example, I = P/V.
Source: Sean White

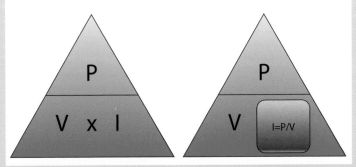

Convert to kWh

2472Wh/1000 Wh per kWh = 2.472 kWh

On a test or in real life, this could be rounded off to 2.5 kWh for the correct answer.

Remember that on the NABCEP tests, there are always 4 choices to choose from and you should make sure that your answer seems right. Remember that all good exam questions have wrong answers that were carefully thought out. Be careful of getting your units right and do not confuse Wh with kWh. You should have plenty of time for the NABCEP Entry Level Exam, so do not rush. Make sure your answer makes sense. I like to take an educated guess at the range of right answers before I complete the problem. Another common mistake is moving the decimal in the wrong direction when converting kWh to Wh. Remember that when converting to large units, your final number will be smaller. Every time you do math, make sure your numbers make sense. Even the smartest people make simple mistakes.

Common mistake:

Energy is kWh not kW/h or kW

BATTERY MATH

When talking about **battery capacity**, the term **Amp Hour** is commonly used. An Ah is an amp for an hour. Amp hours are easily converted to Wh.

$Ah \times V = Wh$

A 6 volt battery that has 100 Ah has:

$6V \times 100Ah = 600Wh$

A 12V battery with 100Ah has:

$12V \times 100Ah = 1200Wh$

An easy way to convert Wh to kWh is to move the decimal 3 places.

kWh are bigger than Wh, so 1200Wh = 1.2 kWh

OHM'S LAW AND RESISTANCE

When power is consumed, we can talk about a relationship between current and voltage called resistance.

The relationship is called **Ohm's law** and is:

V = I × R

This means that when current goes up and resistance remains constant, then voltage also increases. This is called a **proportional relationship**.

When current goes through a wire, there are losses and those losses are consumed by the wire in the form of heat. The losses are known as **voltage drop**. If the current is increased, so is the voltage drop/loss.

WHAT REALLY IS AN OHM?

An **ohm is a unit of resistance** also symbolized with the **Greek symbol Ω**.

Since V = I × R (We also can write V = IR)

Then R = V/I

> Since V was alone, it gets to be on top when I is put on the other side of the equation to solve for R.

So we can say resistance is measured in volts per amp.

For entry level PV, we do not have to go very much deeper into resistance.

Let's do something fun with an inefficient 100W incandescent light bulb.

Practice question:

Using W = VI and given that the voltage of the receptacle is 120Vac, how many Amps will be going through the light bulb?

Answer:

If W = VI and we want to solve for I then

I = W/V
I = 100W/120V
I = 0.833 Amps

Therefore, next time you go to the store, ask for a 0.833A light bulb.

If they don't know what you are talking about, then try asking for your light bulb in Ohms.

Since V = IR then to solve for Resistance

R = V/I
R = 120V/0.833A
R = 144 Ω

So ask for a 144 ohm light bulb.

Using W = VI and V = IR together, we can solve for W, V, I and R if we know any two of the factors. The Ohm's Law Wheel at the beginning of this chapter gives derived equations to help electricians with this.

We could have looked at the Ohm's Law wheel and seen that:

$R = V^2/P$
$R = 120V^2/100W = 144$ ohms

HOW TO USE A METER TO MEASURE VOLTAGE, CURRENT AND MORE

Digital multimeters can typically measure voltage, current and resistance.

Figure 3.3 is an example of a simple and inexpensive meter that can measure voltage, current and resistance.

Figure 3.3 Digital multimeter

Source: Wikimedia Commons. Attribution: oomlout http://commons.wikimedia.org/wiki/File:Digital_Multimeter.jpg. This file is licensed under the Creative Commons Attribution-Share Alike 2.0 Generic license.

VOLTAGE MEASUREMENTS

Always turn the meter to a setting higher than you would ever expect to measure. The meter in Figure 3.3 has a red and a black lead that are used to measure voltage when the connectors are hooked up to the two out of three plugs on the right side of the meter. (When measuring current through the meter, the red lead will go to the left of the black lead.)

When measuring dc (direct current) voltage from PV or a battery, set the voltage to the setting with the solid line over the dashed line.

When measuring utility ac (alternating current), look for the symbol that looks like this ~.

When measuring dc with the meter, the black lead of the meter is negative and the red is positive. In **most PV systems** in the USA, the black dc wire is positive and the **white "grounded conductor" is negative**. That is not a typo, white is most often negative.

When measuring ac voltage, it does not matter if the red and black are one way or the other, since the voltage and current are constantly alternating on each side of zero.

MEASURING CURRENT

When measuring current with a multimeter as pictured in Figure 3.3, the meter will become part of the circuit and all of the current will run through the meter.

Most people in the solar industry would rather not have the current run through the meter for safety and simplicity.

We can also measure current without even touching a wire and this can be done with a clamp.

How to measure current with a clamp:

● Put the clamp around one wire in the circuit.
● Do not put the clamp around two wires in the circuit, such as in an extension chord. Every circuit is a circle with a return path and if you get two parts of a circuit, each with current going opposite directions, they will cancel each other out.

When measuring dc current with a clamp-on ammeter, make sure that you are either **clamping on the negative or the positive, but not both**.

Figure 3.4 Digital multimeter with clamp-on ammeter

Also, most clamp-on ammeters only measure ac current, so when you are shopping for clamp-on ammeters, make sure that you purchase one that has a clamp that works for dc, which is more expensive than an ac clamp.

MEASURING RESISTANCE AND CONTINUITY

Most digital multimeters can measure resistance and a good example of this is to measure continuity. Most meters will beep when there is a connection between negative and positive when they are set to measure continuity (continuity meaning continuous).

Testing a fuse: Set the meter to read Ω and then touch the black and red to each end of the fuse. If it beeps, then there is continuity and the fuse is good. **Make sure that you do not have the fuse installed or energized when you are doing this test.**

HORSEPOWER

Often, pumps and some other electrical equipment are measured in horsepower. It is highly recommended that you know how to convert HP to watts.

1 Horsepower = 746 Watts

A 7.5kW system would be about 10 horsepower.

TYPES OF DC CONNECTIONS (SERIES AND PARALLEL)

Often times PV is hooked up in groups of solar modules where the positive lead of one module is connected to the negative lead of the next module. That is called a series connection. With PV and batteries, **series connections increase voltage**.

Even within a solar PV module itself, the solar cells are connected in series to increase voltage.

Parallel connections do not increase voltage and increase only current with PV.

Series: Positive to negative and increases voltage.

Parallel: All of the positives are connected to one place and all of the negatives to another.

With alternating current, we can also have series and parallel connections.

An ac series connection would be like hooking up those Christmas lights that all go out when one goes out. This is analogous to wiring PV in series and

Figure 3.5 Diagram of a PV system, including series and parallel connections

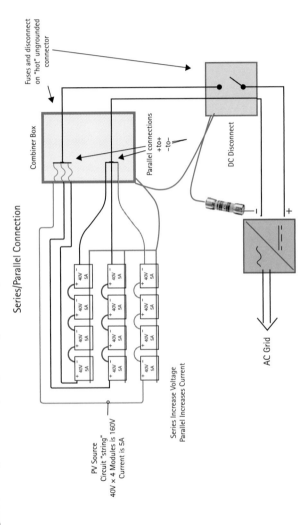

Series/Parallel Connection

Fuses and disconnect on "hot" ungrounded connector

Combiner Box

Parallel connections
+ to +
– to –

DC Disconnect

PV Source
Circuit "string"
40V × 4 Modules is 160V
Current is 5A

Series Increase Voltage
Parallel Increases Current

40V – 5A

AC Grid

In a negative grounded system, inside inverter there is a connection between negative and ground. If more than one amp goes through this connection, there is a ground fault and device deactivates inverter.

when one PV module is shaded, it reduces the current through all of the PV modules. In a series connection, if the current goes through one thing, it then goes through whatever it is in series with.

Statistics for an average Solar PV Module:

● An average solar module is often called a solar panel, but technically it is a solar module.

● An average solar module has 60 solar cells connected in series. You can see the connections going from the front of one cell to the back of the next. (Usually the front is the negative side of the cell.)

● The average solar cell is 6 inches (156 mm) in diameter.

● The average solar module is about 250 Watts (0.25kW) and it takes 4 average solar modules to make one kW.

● The average solar module is slightly less than 40 inches wide and slightly less than 66 inches long.

● A datasheet with the specifications of a solar module can be easily found by searching "250W solar pdf" on the Internet.

Even with water flow in solar hot water systems, we can have series and parallel connections.

An ac parallel connection would be the connections that are made at a subpanel. Various circuits are connected together with a parallel connection.

ELECTRICAL TRANSMISSION AND DISTRIBUTION SYSTEMS

Most often, electricity is made at huge centralized power plants at high voltages by rotating machinery.

The voltage is stepped up for long distance transmission and then stepped down for our use. High voltage has fewer losses.

Figure 3.6 Series and parallel connections. Light bulbs on the right are connected in series to each other. Light bulb in the middle is connected in parallel with the other bulbs.

Why we transmit electricity with high voltage:

Voltage × Current = Power

Higher current causes voltage drop and power loss on transmission lines, so if we can transmit our power with high voltage and low current, then we have fewer losses.

A 1kW 12V battery based system would have 4 times more losses than a 1kW 24V battery based system or 16 times the losses of a 1kW 48V system if we were using the same conductors (wires).

When we lose power on a conductor, it is because of "voltage drop". If we measured the voltage of where power is coming from and where it is going to, we would always have more voltage where the power is coming from. In a PV system, the voltage is coming from the PV to the inverter and coming from the inverter to the grid.

Transmitting electricity over long distances and raising and lowering voltages with transformers always leads to losses. A big benefit of PV is that we can make the electricity where we are using it and do not have to have distribution or transformer losses. Making it where we are using it is called Distributed Generation (DG).

Figure 3.7 Electrical generation, transmission and distribution system

Source: Wikimedia Commons. Attribution: J Messerly https://en.wikipedia.org/wiki/File:Electricity_grid_simple-_North_America.svg.

This file is licensed under the Creative Commons Attribution 3.0 Unported license.

Nikola Tesla first came up with alternating current and transformers. He showed that we could step up and down voltages, so that we can transmit power over long distances with high voltage. Transformers do not work with dc power.

> What is high voltage?
> There are many different definitions of high, medium and low voltage. Utilities think of 600V and below as low voltage. Laypeople think of 600V as high voltage.
>
> People that work with 12V lighting systems think of 12V as low voltage.

Inverter companies sell medium voltage ready inverters and they are referring to medium voltage as thousands of volts (kV).

CHAPTER 3 PV MATH

1. A 200 watt light is left on for 7 days. How much energy is consumed?
   a. 34kWh
   b. 336kW
   c. 3360Wh
   d. 14MWh

2. If a 100W light bulb is working on a 120V socket, what is the current?
   a. 12kW
   b. 0.83A
   c. 1.2A
   d. 144 Ω

3. How much resistance does the light bulb have in question 2 above?
   a. 0.83 ohms
   b. 12 ohms
   c. 0.007 Ω
   d. 144 Ω

4. The hydraulic analogy for voltage is
   a. Flow
   b. Volume
   c. Capacity
   d. Pressure

5. The hydraulic analogy for current is
   a. Flow
   b. Volume
   c. Capacity
   d. Pressure

6. The hydraulic analogy for energy is
   a. Flow
   b. Volume
   c. Laminar
   d. Pressure

7. 2000 watts is equal to
   a. 20kW
   b. 2kW
   c. 0.2kW
   d. 2kWh

8. The power of a 250 watt solar module is equal to
   a. 1/4th of a kW
   b. 0.025MW
   c. 25KVA
   d. 0.25kWh

9. One horsepower equals
   a. 0.746kW
   b. 746kWh
   c. 1000W
   d. 1MW

10. A pump works at 4A and 12V for 3 hours. How much energy does it consume?
    a. 144kWh
    b. 48Wh
    c. 0.144kWh
    d. 96Wh

Answers in back of book.

Solar energy fundamentals

SOLAR POWER

Measured in **power per unit area** and most often in **watts per square meter**.

1000 watts per square meter is called "**peak sun**", which is a typical sunny day near noon.

SOLAR ENERGY

Since: power × time = energy

Then: **solar power × time = solar energy**

If we have **peak sun conditions of 1000 watts per square meter** for one hour, then we have a "**peak sun hour**" or a "**sun hour**".

Different locations have different amounts of solar energy. We often quantify the amount of energy a location has in average **peak sun hours per day**.

Peak sun hours per day is also often called insolation. **Insolation is incident solar radiation**.

Table 4.1 gives some insolation data for various locations when the solar collector is tilted at a latitude tilt angle. (Latitude is the angular degrees from the equator, which we will learn more about later in this chapter.)

PSH = Peak Sun Hour = 1kWh/m^2/day at latitude tilt.

Table 4.1 Insolation data

Location	Daily Insolation
San Francisco	5.4 PSH
Las Vegas	6.5 PSH
Seattle	3.7 PSH
Honolulu	5.7 PSH
Guam	5.1 PSH
New York City	4.6 PSH
Chicago	4.4 PSH

Insolation data from NREL Redbook

http://rredc.nrel.gov/solar/pubs/redbook/

Standard Test Conditions:

- **1000 watts per square meter**
- **25°C (77°F)**
- **1.5 Atmospheric Mass**

Standard Test Conditions (STC) are **how PV modules are tested** and are the conditions under which the module performance is measured. These are the statistics, which are put on the label and determine how many watts the module has, which determines the price. Sometimes PV will make more power, current and voltage than what it says on the label (STC).

Four 250W PV modules make 1kW of PV.
Forty 250W PV modules make 10kW of PV.
Four thousand 250W PV modules make 1MW of PV.

Many times it is stated as **10kWp** where the p stands for peak sun conditions, which is also STC.

The temperature will determine the voltage. Cooler PV works better and has better voltage. **Lower temperatures increase PV voltage.**

Figure 4.1 The atmosphere filters more light at greater angles

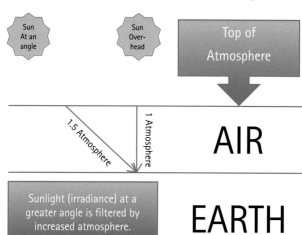

1.5 Atmospheric Mass (AM) is the spectrum of light, which is tested in the factory. The test conditions are simulated with a flash of light and a filter to simulate 1.5 AM during PV module testing. The thicker the atmosphere, the less light will make it to the PV. This means that at higher elevations, sunlight will have less atmosphere to travel through, which will increase performance. Also, around noon and summer solstice, the sun is highest in the sky and will have less atmosphere to travel through, than when the sun is lower in the sky.

> The atmosphere is relatively thin compared to the earth. If we painted over a globe, the thickness of the paint would be about that of the atmosphere. If we could drive our car straight up for five minutes at highway speeds, we would reach the edge of the atmosphere. That is not a lot of room to dump 100s of years of CO_2 pollution.

LATITUDE

The equator is zero latitude and the North Pole is 90 degrees latitude. Just like degrees in a compass, the equator is at a 90 degree angle from the North

Pole. At **every latitude, the path of the sun is exactly the same all around the world**. Athens Greece and San Francisco California are both at latitude 38 degrees and have the same sun paths. The weather will make solar energy production different however.

SOLAR TIME

In the early 1800s, before there were time zones, we kept time by where the sun was in the sky. **Solar noon** is when the sun is as high in the sky as it will get for the day. In the latitudes of the continental United States and Canada, **solar noon is when the sun is due south**. (In the tropics and Southern Hemisphere, the sun can be north at solar noon.)

As we can see from Figure 4.2, the earth rotates around the sun in the **ecliptic plane**. Most everything in the solar system rotates around the sun in this same plane and spins in the same direction.

From a northern hemisphere perspective (where most people are), everything spins counter-clockwise. This is why it is noon in NYC three hours before it is noon in San Francisco.

The earth is tilted 23.5 degrees, this is what gives us seasons. The equator is facing 23.5 degrees below the sun on summer solstice and is facing 23.5 degrees above the sun on winter solstice.

At equinox, the sun will be right over the equator.

> Equinox is Latin for equal nights. At equinox when the sun is right over the equator, the sun will rise at 6am solar time and set at 6pm solar time everywhere in the world (except the north and south poles). Equinox is the only time of the year when the sun rises due east and sets due west. In the northern hemisphere, the sun will rise and set to the north of east and west in the summer and rise and set to the south of due east and west in the winter. At summer solstice at sunrise and sunset, the sun will be more north than south and will shine on a north-facing window.

Figure 4.2 Earth's orbit

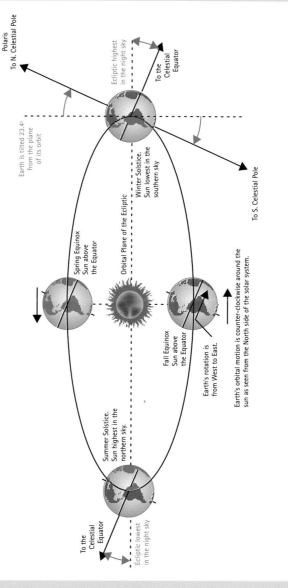

If we want to see the path of the sun plotted on a sun chart, we can do an Internet search for "Oregon Sun Chart" and get a sun chart from the University of Oregon.

By using the sun chart in Figure 4.3, we can predict where the sun will be throughout the year.

Figure 4.3 Image generated from University of Oregon Sun Chart Program

On the horizontal x-axis azimuth is plotted.

On the vertical y-axis the elevation of the sun is plotted.

> The sun is the highest when it is over the meridian or longitude of where you are located. Meridians or lines of longitude are lines that indicate where a location is in the east and west direction. These lines go from the north to the south poles along the surface of the earth. Meridians or lines of longitude are perpendicular to lines of latitude.

Azimuth is the direction we measure on a compass. North is zero degrees and we go clockwise 360 degrees. East is 90 degrees, south is 180 degrees and west is 270 degrees azimuth.

(Occasionally in the solar industry, we see some systems that use south as zero degrees azimuth).

Figure 4.4 Azimut altitude.svg
Source: Wikimedia Commons. Attribution: Joshua Cesa. http://commons. wikimedia.org/wiki/File:Azimut_altitude.svg. This file is licensed under the Creative Commons Attribution 3.0 Unported license.

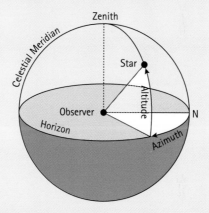

In Figure 4.5, we have a compass with a correction for magnetic declination, such as we see on the West Coast of the United States. The red arrow of the compass is pointing at the magnetic North Pole. The adjustable dial has been rotated counter clockwise towards the geographic North Pole (the axis of the Earth). In the solar industry, we are looking for the path of the sun, which has to do with the axis of the earth. We usually only have to correct for magnetic declination when using a magnetic compass. GPS, maps and many solar devices already correct for magnetic declination. Remember that the magnetic North Pole is over the middle of Canada.

Figure 4.5 Compass with a correction for magnetic declination
Source: Courtesy of Blake Miller, http://outdoorquest.blogspot.com/2011_
09_01_archive.html.

Figure 4.6 Map for correction of magnetic declination

Figure 4.6 shows how we correct for magnetic declination in different locations. For instance in Florida, we subtract 5 degrees from the azimuth as read on a magnetic compass to correct for magnetic declination.

Figure 4.7 Magnetic North Pole moves slowly
Source: Wikimedia Commons. Attribution: Tentotwo. https://en.wikipedia.org/wiki/File:Magnetic_North_Pole_Positions.svg. This file is licensed under the Creative Commons Attribution-Share Alike 3.0 Unported license.

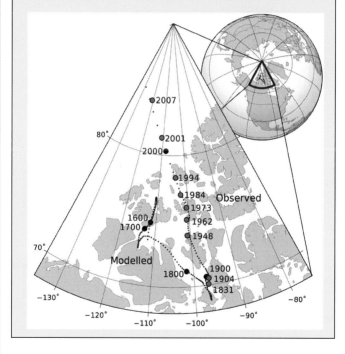

Solar noon by definition is when the sun is highest.

Get to know the sun chart and you can use it to predict shading of a solar array.

You can also be an expert on predicting the best places for sunsets.

If we look more closely at the sun chart, we see that hours are marked with solar noon being when the sun is south. The other hours of the day are also marked. In the months between March and September, the days are longer than the nights and in the months between September and March, the nights are longer than the days.

At equinox, we can see that the sun rises at 6am solar time and sets 12 hours later at 6pm solar time. This is true for all sun charts.

The days closest to the shortest day of the year, which is winter solstice and is approximately December 21, will have the shortest shadow.

The seasons do not always start on the same day throughout the year. This is because of leap years, and earth's orbit that is not perfectly symmetrical (it is elliptical). Also, in different places on the earth, it is different days at the same time. There is a good chance when you are reading this, that it is tomorrow in Asia.

We can use the sun charts to help determine how much solar energy will be at a certain location and there are devices, which can superimpose trees and other objects that cast shadows onto different sun charts.

The two most popular devices are the Solmetric Suneye and the Solar Pathfinder.

With the Solar Pathfinder and the Solmetric Suneye, summer is towards the middle, which corresponds closest to overhead.

On the University of Oregon sun chart, summer is plotted with a higher solar elevation angle over the top.

THE TROPICS

Below 23.5 degrees north latitude, the sun will be directly overhead at times. The **23.5 degrees north latitude line is called the Tropic of Cancer**.

Above 23.5 degrees south latitude will also have the sun overhead at times and this line is called the Tropic of Capricorn (in the southern hemisphere).

Figure 4.8 The Solmetric Suneye is a popular device used for shade analysis and can predict the path of the sun throughout the year.
Source: www.solmetric.com

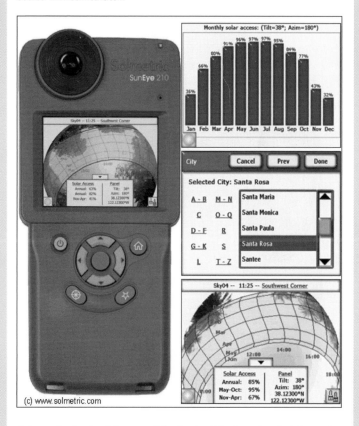

(c) www.solmetric.com

Between the Tropic of Cancer and the Tropic of Capricorn is called the tropics. The tropics is the only place where the sun can be overhead.

In the north where the sun never rises at winter solstice nor sets at summer solstice, it is called the arctic and the arctic begins at the Arctic Circle, which is

Figure 4.9a and b The Solar Pathfinder is another popular device used to predict shading and the path of the sun throughout the year.

Source: www.solarpathfinder.com

(a)

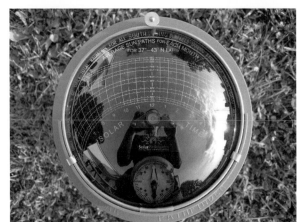

(b)

at 66.5 degrees latitude. This has to do with the tilt of the earth, which is 23.5 degrees latitude (90°-23.5°=66.5°).

The same can be said for the Antarctic Circle, which is at 66.5 degrees south latitude.

Figures 4.10 and 4.11 Angles of the sunlight hitting the earth at winter solstice and summer solstice. By looking at these images, we can see how the sunrays hit the earth at different times of the year.

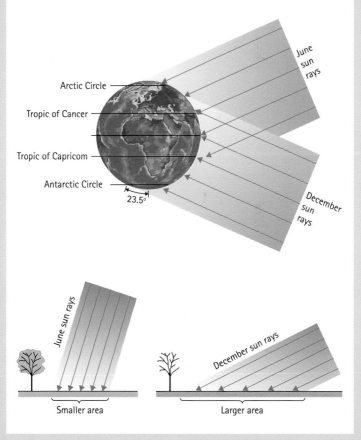

Figure 4.12 The sun paths as viewed from the earth. This is what the sun path looks like in areas that are not in the tropics or the arctic, which is most of the United States and Canada. In the tropics, the sun will be overhead sometimes and in the arctic, it will not rise at the dark of winter.

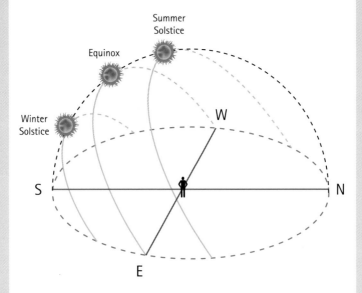

TILT ANGLES

The sun constantly is moving throughout the sky from our perspectives. It will go 360 degrees in 24 hours, which is 15 degrees per hour and 1 degree every 4 minutes.

When the sun is on the horizon, the elevation angle would be zero degrees. In order for PV to directly face the sun, the PV would have to have a tilt angle of 90 degrees when the sun is on the horizon.

If the sun would be overhead, the optimal tilt angle would be zero degrees.

Tilt of PV is based on degrees from the horizontal. Flat is zero tilt and vertical is 90 degrees tilt.

In order to maximize energy production some would have a tracker that would follow the sun. A **dual axis tracker should always face the sun** and get maximum energy and a **single axis tracker will track the sun on one axis**.

Concentrating PV systems need to use 2-axis trackers to focus direct beam solar radiation.

Figure 4.13 A horizontal single axis tracker faces east in the morning and west in the afternoon (flat at solar noon).
Source: Wikimedia Commons. Attribution: Msrt10. http://en.wikipedia.org/wiki/File:RayTracker_Utility_Scale_Solar_Tracker_Installation.JPG. This file is licensed under the Creative Commons Attribution-Share Alike 3.0 Unported license.

Figure 4.14 2-axis tracker with concentrating PV always faces the sun
(Tesla Roadster EV charging)
Source: Wikimedia Commons. Attribution: Mbudzi. http://commons.wikimedia.
org/wiki/File:Amonix7700.jpg. This file is licensed under the Creative Commons
Attribution-Share Alike 3.0 Unported license.

Most often, trackers are not used since they add moving parts and complexity.
Depending on where you are and the type of tracker, a tracker can harvest 20
to 30% more energy than a system without a tracker. As the price of PV drops,
trackers are used less.

Usually on a residential roof, PV is flush mounted to the roof and is put a few
inches above the roof facing exactly the way the roof faces.

The US government sponsored National Renewable Energy Laboratory
(NREL) has a website that will help you determine energy production.
The popular Internet tool is called PV Watts. Do an Internet search for
PV Watts for simple PV system production modeling. NREL has many
resources for renewable energy education and research. When designing a PV system, it is very important to know how much energy your
system will make.

Figure 4.15 Tilt angle optimization for different seasons

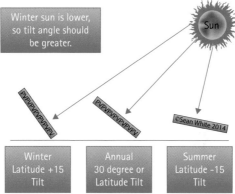

We usually try to get the solar modules facing the direction where they will collect the most energy. There are some "rules of thumb" that we use in the solar industry. They are not exact, but are good to know (often on exams).

Solar rules of thumb:

1. Keep your PV array shade free from 9am to 3pm
2. Optimize for **summer** production with a tilt of **latitude minus 15 degrees**
3. Best for **winter** production tilt of **latitude plus 15 degrees**
4. **Best for annual** production is either latitude tilt or **30 degree** tilt depending on the location.

SHADING

Shading of a PV system is to be avoided, especially when the sun is high in the sky.

Since solar is often wired in series, shading one solar cell can affect the other cells that it is wired in series with. This is often compared to the Christmas light effect: when one light is removed, it will affect the other lights it is wired in series with.

INTER-ROW SHADING

When PV is arranged in multiple rows, the row to the south can shade the row to the north. There are many factors that will help determine how far the rows are to be spaced. The tilt angle, the size of the PV and the height of the back of one row of PV compared to the front of the next row along with the latitude will determine inter-row spacing.

Notice the shadow on the row of modules to the left. This is unacceptable unless it is early or late in the day.

In lower latitudes, such as in Miami Florida, rows can be spaced closer together than in places that are farther north, such as in Toronto Canada. This is because places that are farther from the equator have a sun path that is lower in the sky, especially in winter.

Figure 4.16 M&M candy factory in NJ
Source: Photo by Sean White

Figure 4.17 Inter-row shading
Source: Image by Sean White

Inter-row Shading

Inter-row Spacing can be determined with trigonometry or with rules of thumb, such as 2:1, 2.5:1 or a 3:1 ratio of distance to height.

Figure 4.17 demonstrates the height of the row and the distance from the back of one row to the front of the next. Often times we talk about a ratio of height to distance of the shading object and the distance away from that object. Close to the tropics, 2:1 distance:height is acceptable and farther north 3:1 is more acceptable.

Some people use trigonometry (the relationships of triangles) to determine the best inter-row spacing.

Many times commercial rooftop systems have a lower tilt angle in order to decrease the height of the rows and fit more PV on the roof. Also lowering the tilt will cause less wind uplift forces.

TYPES OF SOLAR RADIATION

Direct: Radiation that comes directly from the sun. The rays of sunlight are parallel to each other.

Diffuse: Radiation that comes from bouncing off of clouds and the atmosphere.

Albedo: Radiation that is reflected off something else, such as the ground, a white roof, a tree or a lake.

Global: Combined solar radiation from all types.

> Concentrated PV is solar energy, which is captured with lenses or mirrors that focus sunbeams onto solar cells. Concentrated solar only works with direct sunlight and does not work at all on a cloudy day. Concentrated PV only works with a 2-axis tracker.

WAYS TO MEASURE SOLAR RADIATION:

Pyranometer: A device that measures global solar radiation. Also called an **irradiance meter**. Pyro = fire and meter = measure. Measures global solar radiation.

Pyranoheliometer: Measures **direct beam** solar radiation. Direct beam solar radiation is important for concentrating systems.

The sun is constantly moving throughout the sky and weather patterns in the atmosphere make solar energy analysis a science that can be as variable as the weather on a daily basis and as variable as the changing climate on an annual basis.

Most residential rooftop PV systems are not oriented in the perfect direction, but they do produce enough solar energy to be cost effective. As solar systems go down in price, new rooftop space that was not cost effective in the past becomes cost effective. In places like Hawaii where energy is expensive and sunshine is high in the sky, even north facing arrays are now becoming good investments.

PV module fundamentals

Albert Einstein won the Nobel Peace Price for describing the process by which light interacts with matter and "sets electrons free". He was in the solar industry, just like you, which is smart.

HOW PV WORKS

Light (photons) hits a semiconductor, increases the energy of electrons and creates current, voltage and power.

Figure 5.1 P-N junction of a photovoltaic cell
Source: Sean White

Figure 5.2 Monocrystalline solar cells, rounded corners
Source: Courtesy Suntech

Silicon is refined from sand or quartz, and then formed into a large crystal ingot, which is then sliced into thin wafers. Wafers are then prepared by adding different layers of phosphorous, anti-reflective coatings and screen-printed silver contacts for catching electrons. There are different variations on this theme, but this is your typical process.

Nominal Voltage: Nominal means "in name only". The nominal voltage of your car battery is 12V (2 volts per battery cell). The nominal voltage of a silicon solar cell is one third of a volt. 36 cell PV modules were popular in the past, because they were 12V nominal (36 × one third = 12V). Before grid tied solar was popular, we would match PV voltage to battery voltage at a ratio of 6 PV cells to each lead acid battery cell.

Solar cells are combined to create a solar module. An average solar module consists of 60 solar cells connected together in series.

Figure 5.3 Polycrystalline/multicrystalline solar module
Source: Courtesy Yingli Solar

Multicrystalline solar cells are cast cells with square corners.

Monocrystalline is typically more efficient than polycrystalline solar, however the differences are not usually very significant, especially since monocrystalline is usually more expensive and has a higher carbon footprint.

The **most efficient solar cells** are made from a material called **Gallium Arsenide**. This is very expensive and usually used for outer space and concentrating PV applications.

Other types of thin film solar are

1. Cadmium Telluride
2. CIGS Copper Indium Gallium Selenide
3. Amorphous Silicon.

Thin film is not as popular, because it is typically not as efficient and cost effective as crystalline silicon PV.

SOLAR PANEL

Most people call a solar module a solar panel.

Figure 5.4 Solar panel made from solar modules
Source: Courtesy Sunlink

Figure 5.5 A PV array is a larger group of modules.
Source: Wikimedia Commons. https://commons.wikimedia.org/wiki/File:Vermont_
Law_School_Solar_Panel_Array-1.JPG. This file is made available under the Creative
Commons CC0 1.0 Universal Public Domain Dedication.

Technically, a solar panel is a group of modules preconfigured as a unit before they are installed.

Calculating efficiency of a solar module:

Efficiency is the ratio of watts per square meter of output at STC to the irradiance at STC, which is 1000 watts per square meter.

For instance, if a PV module was making 150 watts per square meter, then 150 watts per square meter output / 1000 watts per square meter (STC) input = .15 or 15% efficiency.

Let us do an example calculation:

Example 250W PV Module
Dimensions 1638 × 982mm

1.638m × 0.982m = 1.6 square meters
250W/1.6 square meters = 156W/square meter

63

Figure 5.6 Typical solar module dimensions in mm
Source: Courtesy Canadian Solar

1638.0

982,0

STC is 1000 watts per square meter

156W/1000W = 0.156

0.156 × 100% = 15.6% efficient PV

ESTIMATING HOW MUCH PV FITS IN A GIVEN SPACE

If you covered 100 square meters with 15.6% efficient PV, you would have

156W/square meter × 100 square meters =
15,600W = 15.6kW of PV

In reality, you would need extra space around the edges and chimneys, since you usually cannot fit PV in every space.

Some designers estimate 100W to 120W/square meter, to leave room for roof penetrations, walkways and space for firefighters.

Do not use dissimilar PV modules in series.

When PV is connected in series, you will have what is sometimes called "the Christmas light effect". **If one module has less current, then it will limit the amount of current to flow through the whole series connected PV source circuit (string).**

Before there were bypass diodes in PV modules, when a cell was shaded all of the current from all of the other cells and modules would try to push through the shaded cell. This would create heat (from resistance) and could be a fire hazard. Also it would be terrible for performance.

Figure 5.7 Module Mismatch

Module Mismatch is Bad

+ [3A] – + [5A] – + [2A] – + [4A] –

Hook up these modules
in series and you get 2A

This is similar to what happens with shading or
different orientations within a string

Figure 5.8 PV module with bypass diodes
Source: Courtesy of Solmetric

PV Module with Bypass Diodes
Current can bypass groups of shaded cells and sacrifice voltage to get
through. Typical PV has 3 sections and will skp 1, 2 or all 3 at once.

Shading of the long edge is more desirable than shading of
the short edge due to the arrangement of the diodes.
Shading a row on the short edge will take out all 3 sections.

Bypass diodes will bypass a group of cells. The group of cells being bypassed
would not create voltage, but would let the current through.

Bypass diodes sacrifice voltage for current.

A bypass diodes analogy would be: When there is an accident on the freeway,
you take the frontage road.

Bypass diodes are **wired in parallel** to a group of cells so there is another
pathway for current to take when there is a shaded cell.

BLOCKING DIODES

Blocking diodes are not in the typical PV system these days, however in the
past they were popular. They **prevent reverse current**. In plumbing an analogy
would be a check valve. Another highway analogy is a one-way street.

Blocking diodes are wired in series.

IV CURVES (I = CURRENT AND V = VOLTAGE)

(Good section to bookmark and review)

Voc = OPEN CIRCUIT VOLTAGE

In Figure 5.9 we can see the point called Voc (open circuit voltage). Voc is the point at which there is an open circuit. This means that the **PV system is turned off**. As we can see, the voltage is as high as it can get, but the current is zero. Since voltage multiplied by current is power, then:

Voc × Zero Amps = Zero power
(Amps is current is I)

It is interesting that when you turn off a PV system, the voltage will increase. That means you can still get shocked when PV is off.

ISC = SHORT CIRCUIT CURRENT

Short circuit current (Isc) is the upper left point on the IV curve where there is **all current and no voltage**. This is an operating point that hopefully your PV system will never experience. A short happens when there is a **direct**

Figure 5.9 I-V curve
Source: Courtesy Solmetric

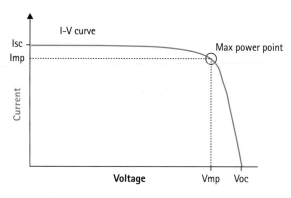

connection between positive and negative. This is the point of maximum current, where voltage is zero. Once again anything multiplied by zero is zero, so Isc is not a power-producing situation.

Isc × Zero Volts = Zero Power

If conductors are installed wrongly and the insulation of the wire disappears, you will have a **positive to negative short**. This can cause a fuse in a combiner box to activate.

Vmp = MAXIMUM POWER VOLTAGE

At maximum power voltage we have the point of maximum power. This means that we want to operate in this position to get the most out of our PV. It is interesting to note that "**maximum**" power voltage **is less than** open circuit voltage.

Imp = MAXIMUM POWER CURRENT

Imp is maximum power current and this is at the same point as Vmp.

If we multiply Imp × Vmp we will get the maximum power at the maximum power point.

Maximum power current is less than short circuit current.

Vmp × Imp = Maximum Power

For example, if we have a PV module where the Vmp is 30 Volts and the Imp is 8 Amps then:

30 Vmp × 8 Amp = 240W

We have a 240W PV module

Often students get confused and think that Vmp or Imp is the higher voltage or current, since m is for maximum, however it is not the case. Isc is higher current than Imp, but at Isc there is no power, not maximum power. Isc is not a power-producing place, and the same goes for Voc.

Another common **mistake** people make is **multiplying Voc and Isc** and getting some imaginary power. This does not get you power!

If we look at Isc at the upper left part of the IV curve, and see how the line gradually slopes down to the maximum power point (MPP) we see places where power is made, but not as much as the power made at the MPP. It is interesting that **at the point of the most current, there is no power at all**.

Most inverters can keep the PV working at the MPP and we call this maximum power point tracking or MPPT.

Summary:

Voc × Zero Amps = Zero Power
Isc × Zero Volts = Zero Power
Vmp × Imp = Maximum Power
Voc × Isc: PV cannot operate here

FACTORS THAT CHANGE PERFORMANCE OF PV

When PV is made and sold it is rated at STC (Standard Test Conditions)

STC
1000W/m^2 (brightness of light about like noon)
25°C   (77°F)
1.5AM   (Atmosphere Mass spectrum of light)

We need to be able to compare PV modules and determine which is the best deal. When we pay a per watt price, we are getting the PV rated at STC.

In the real world, PV operates under different conditions.

The **best production we can get is on a bright cold day**. Mt. Everest may be a good place for PV!

TEMPERATURE AND VOLTAGE (ENVIRONMENTAL CONDITIONS)

Figure 5.10 Temperature and irradiance effects on voltage and current
Source: Sean White

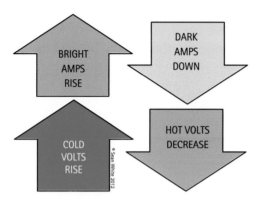

BRIGHT = MORE CURRENT

Brightness means more light (photons) knocking electrons loose, which translates to more current (amps). We call bright light increased irradiance and since irradiance starts with the letter I, then it is easy to remember that it increases I (symbol for current).

COLD = MORE VOLTAGE

Cold temperatures increase voltage. Some students think of the T in volt to remember.

Silicon semiconductors just work better in the cold. That is why your computer has a fan that cools the silicon computer chip.

The problem with cold temperatures increasing voltage is that sunlight, which is good for solar cells, causes heat, which is bad for power production.

Figure 5.11 Irradiance and temperature effects on I-V curves
Source: Courtesy Canadian Solar

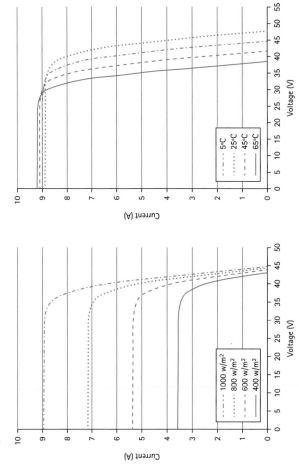

On the left of Figure 5.11 are IV curves with different amounts of irradiance (light) with a constant temperature. We can see how **more irradiance increases I (current)**.

On the image on the right of Figure 5.11, we can see that when the temperature is changed voltage changes. **Colder temperatures cause higher voltages**. That is why we like to mount our PV so that there is **airflow, to blow the heat away**. PV can get very hot in the sun. But it is better to have a hot sunny day than a cold dark day, because with darkness, you lose more power than with heat.

In Figure 5.12, we can see how PV stacks up with series and parallel connections.

This image represents **two strings (PV source circuits) of three modules in series**.

From this image, we can see how voltage increases with series connections. On a cold day, this voltage will increase even more.

Figure 5.12 I-V curve representation 2 PV source circuits (strings) of 3 in series
Source: Courtesy of Solmetric

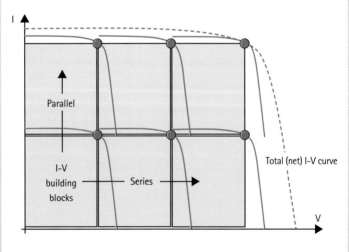

DIFFERENT WAYS OF TESTING PV

Table 5.1 PV testing conditions

STC	NOCT	CEC	PTC
25°C cell	20°C ambient	20°C ambient	20°C ambient
1000W/m2	800W/m^2	1000W/m^2	1000W/m^2
	1m/sec wind	1m/sec wind	1m/sec wind

STC, Standard Test Conditions is measured in watts and is the standard way modules are tested and sold. STC is what is always indicated on the back of the module.

NOCT, Nominal Operating Cell Temperature is the temperature that the PV solar cell heats up to when exposed to NOCT test conditions.

CEC (California Energy Commission) and **PTC** (Performance Test Conditions) are similar tests, which are extrapolated with different formulae. This is a lower power than STC because of the solar cells heating up in the light.

AM (Atmospheric Mass) 1.5

All of the tests are carried out with the spectrum of light equivalent to **1.5 atmospheres**. By definition 1 atmosphere is when the sun is directly overhead at sea level. This test was standardized when the atmosphere was measured at Cape Canaveral Florida on Equinox.

PV module label data requirements:

1. STC Watts
2. Voc
3. Vmp
4. Imp
5. Isc
6. Maximum series fuse rating
7. STC conditions

Not required on PV label:

1. CEC (California Energy Commission)
2. NOCT (Nominal Operating Cell Temperature)
3. Minimum fuse rating (no such thing).

Explanations:

Maximum series fuse rating is on the PV module, because if you had a fuse that was too big (too much current before opening the circuit) then it would not protect the module assembly. We say the module assembly includes the connectors and everything else that comes with the module. Over 90% of modules have a 15A max series fuse rating. If you put a 30A fuse on it, then it would not protect the module and that would be a safety issue. If you put a 5A fuse on it, it would protect it, but you would have blown fuses opening circuits when you didn't need to, which is not a safety issue.

PV AS A LIMITED CURRENT SOURCE

PV Isc is only about 7% more than Imp. Usually when electricians are working with conventional electricity, they have a current source that can be much greater than the operating current. There are benefits and drawbacks to this.

When we pull current out of the wall in our house, there is a constant voltage source and we take as much current as we need. When there is a short, too much current causes a circuit breaker or fuse, aka, Over Current Protection Device (OCPD) to turn the system off by opening the circuit.

With PV, since the short circuit current is not much more than maximum power current a fuse will not always open the circuit when there is a short. This is the drawback.

A benefit is that there is not too much extra available current unlike with conventional power.

"Listed" PV modules must be used in the United States, Canada and most of the developed world. The PV modules in North America must be listed to the **UL Standard 1703**. This means that there are standards developed by UL and that a testing laboratory must test the PV to that standard. (UL 1741 is for inverters.)

When PV is used to charge batteries without an MPPT charge controller, the PV voltage will have to match the battery voltage. 36 PV cells will work best for charging a 12V battery at about 14.5 volts.

System components (definitions)

Chapter 6

SYSTEM COMPONENTS

PV module: Turns light into electricity. A group of solar cells in a sealed unit.

Grid-tied inverter: Also called **interactive inverter**. An inverter that reads the voltage of the grid and produces as much power and current as it can, regardless of loads. For instance a 5kW inverter can only put out 5kW. Grid tied inverters are sold according to maximum output power. They also have dc and ac operating voltage ranges.

Microinverter: A small inverter that is mounted behind a PV module.

AC module: A module with ac output listed for such use.

Loads: Devices that use energy.

Conductors: Wires.

Conductor sizes: From small (larger numbers) to larger (smaller numbers). 18AWG is a small wire, 0AWG is a large wire and 4/0 (0000) is larger than 0. In order of increasing size wire sizes commonly used from **small to large are 18AWG, 16AWG, 14AWG, 12AWG, 10AWG, 12AWG, 10AWG, 8AWG, 6AWG, 4AWG, 2AWG, 1AWG, 0AWG, 2/0AWG, 3/0AWG, 4/0AWG**.

AWG: American Wire Gauge.

Grounded conductor: A conductor that is intentionally referenced to ground, but carries current. The color of the grounded conductor is **white**. In ac wiring the grounded conductor is also called a neutral. Most PV systems installed in the past in the US are negatively grounded and the negative is white, while the ungrounded positive is usually black. This is not a typo: **the black wire on a**

Figure 6.1 Grid-tied system components

Grid tied system components DC wiring

| PV Source Circuit(s) | | Combiner Box | | DC disconnect | | Inverter | | AC disconnect |

PV source circuit conductors

PV output circuit conductors

inverter input circuit conductors

inverter output circuit conductors

*ac disconnect is often backfed circuit breaker. AC wiring unique to various connection point locations.

negatively grounded PV system is positive and the white wire is negative. The black wire can also be red to avoid confusion. The rule is that the grounded conductor must be white or gray.

Grounding conductor (equipment grounding conductor or EGC): A conductor that is not meant to carry current and connects (bonds) metal equipment together, so that there are no dangerous voltage differences. A grounding conductor is **green or a bare** wire. Often people remember that a grounding conductor is green or bare, because "grounding" ends with the letter G. Even **"ungrounded systems" have equipment grounding**.

Branch circuit: A circuit coming from a circuit breaker off of a load center. Inverters have dedicated branch circuits.

Transformer: A device used for changing one ac voltage to another ac voltage.

Transformerless inverter: Also called **ungrounded inverter** or **non-isolated inverter**. This inverter does not have a current carrying conductor that is at the same voltage as ground. This is the most popular inverter in the world and more common outside of the US, although more common in the US recently. **Transformerless inverters cannot have a white dc wire.** Transformerless inverter exposed source circuits (strings) may **only** be **PV wire** and **not USE-2**.

PV source circuits (strings): PV modules connected together in series including the PV modules and the conductors. Voltage increases in source circuits and current does NOT increase as a result of series connections.

String sizing: Determining the number of PV modules to be connected in series so that voltage is not too high when it is cold (causes over voltage) and voltage is not too low on hot days (under voltage).

PV source circuit conductors: Connects the PV to the combiner box. If there is just one string, there is no need for a combiner (nothing to combine) and the PV source circuit conductors will connect the PV to the dc disconnect.

USE-2 wire: Common wiring method for PV source circuits where conductors are exposed to sunlight or not protected in conduit. Can be used exposed or in "free air".

Conduit: Pipe that is used for protecting wires.

Raceway: A place where wires are protected. Conduit is a type of a raceway.

PV wire: Type of conductor commonly used with PV systems in the same way USE-2 is used. Unlike USE-2 it is the only type of conductor allowed to be **used with transformerless inverters when not in conduit on a roof**. PV wire is usually used at PV arrays with wire that is exposed to the elements. PV wire is better than USE-2 wire.

Combiner box: Where **PV source circuits are combined to become a PV output circuit**. Combiner boxes usually contain fuses when three or more PV source circuits are combined. Combiner boxes are **where parallel connections are made**. **Parallel connections increase current and do NOT increase voltage**. A blown fuse in a combiner can mean a positive to negative short in a PV source circuit.

Junction box: An electrical box where connections are made.

PV output circuit: At the output of the combiner is the PV output circuit. The current of the PV output circuit is the combined current of the PV source circuits going into the combiner box.

Ground fault protection device: In grid tied systems it is **part of the inverter** and will turn the system off if there is a **dc ground fault**. Often called GFDI for ground fault detection and interruption.

DC disconnect: A switch that **disconnects (opens the circuit)** of dc conductors. DC disconnects are most often found near the inverter and near combiner boxes. **PV dc disconnects are not allowed in bathrooms**.

AC disconnect: A switch that disconnects (turns off) ac circuit conductors. A circuit breaker is a disconnect and also an over current protection device.

Over Current Protection Device (OCPD): Protects equipment and conductors from overcurrents. OCPDs will open the circuit if the current gets high enough. PV module source circuits in combiner boxes are usually protected by 15A fuses, which is the typical PV **maximum series fuse rating**. The two types of OCPDs are fuses or circuit breakers.

Circuit breaker: An OCPD and a disconnect all in one.

Service panel: Also called **panelboard** or **load center**. Where circuit breakers are connected to the electrical system. The main service panel is the main one in the building. Other sub-panels are connected to the main panel via feeders. Breakers are connected on busbars.

Busbar: A piece of metal where different circuits are connected in parallel. Usually where breakers or wires are connected.

OFF-GRID SPECIFIC SYSTEM COMPONENTS

Battery: Stores energy by chemical reactions. Most stand-alone PV systems use lead acid batteries.

Flooded lead acid battery: A battery that has a liquid electrolyte (fluid) and uses chemical reactions between lead and acid to store energy.

Sealed valve regulated flooded lead acid battery: A battery that is often called maintenance free, since **you cannot add fluids to a sealed battery**. The valves are for releasing gasses, so the battery does not explode.

Absorbed glass matt and gel batteries: Batteries where the material inside is not fluid. Also considered maintenance free.

Charge controller: Controller that regulates charging of batteries, **so they are not overcharged or undercharged**. Controllers can be for stand-alone systems or grid tied battery backup systems. The **highest voltage** for a charge controller is for the **equalization charge**, which is only used on flooded lead acid batteries that fluid can be added to. The equalization charge is for battery maintenance and only done on flooded lead acid batteries that are **not sealed**.

Battery inverter: An inverter that makes voltage from a battery and produces current and power, as it is needed. A battery inverter is hooked up to a battery and not the PV.

Battery inverter input circuit: The conductors between the battery and the battery inverter, which are **sized according to** the power of the inverter and the **lowest battery voltage**. A lower voltage requires more current for the same power since **power = voltage × current**.

DC coupled battery PV system: A typical off-grid PV system where charge controllers are used to charge batteries.

AC coupled battery PV system: A PV system where **interactive inverters** (grid tied) are used with **battery inverters** to charge batteries.

Grid tied battery backup PV system: A system that **works with or without the utility**. Inverters that can operate in both modes are called **bimodal** or **multimodal inverters**.

Maximum Power Point Tracking (MPPT) charge controllers: Grid tied inverters are MPPT and many charge controllers are MPPT. MPPT is only for devices that are connected to PV and will optimize production. MPPT controls the IV parameters and will **cause the inverter or charge controller to work on the point of maximum power on the IV curve**. An off-grid inverter connected to a battery does not have MPPT.

PV system sizing principles

To size a PV system one must know how much sunlight comes into the system and how much energy is needed to come out of the system, taking into consideration the price of the system.

There are many ways to size PV systems, but the big differences are between grid tied utility interactive and stand-alone off-grid systems.

Stand-alone systems are made to work alone and will have to be sized so that they work during the worst conditions of the year, which usually means getting through a dark December. The month with the worst solar irradiation we call the **critical design month**. In the summer time, there is usually extra energy that could have been used, but is not produced, because it is not needed in an off-grid system. Also, stand-alone systems have extra energy loss due to inefficiencies with charging and discharging batteries.

With a **grid tied system**, in almost all situations, every ac kWh (unit of energy) is used and if the customer is not using the energy onsite, it can be exported to the grid so that others may use the energy. With net-metering, the producer of the electricity will get credit for exported energy and be able to use it months later.

There are always customers who ask to use batteries and be independent from the grid, when they are connected to the grid. Financially, this is not a good investment. Batteries are expensive and customers will not be able to export energy and get credit for it later.

Some customers want to have battery backed up systems when they are connected to the grid. This is not a common practice, because it is a complicated and expensive installation that most PV installers try to discourage their customers from implementing.

BASIC SYSTEM SIZING FACTORS

Grid tied system sizing (without batteries):

1. Space available for system
2. Energy production goal in kWh/yr
3. Customer budget
4. Incentives.

Off-grid system sizing:

1. Energy requirements
2. Power requirements
3. Reducing loads (efficiency)
4. Battery budget
5. PV budget
6. Days of autonomy.

Regardless if the system is utility interactive or stand-alone, the location and derating factors also are a factor in determining system size.

To simplify system sizing, we can look to this basic grid tied example and go from there:

AC kW system size × derating × insolation = production

Example: PV system produces **4kW** at 1000W per square meter of irradiance in a location that **had 5 sun hours** per day of insolation and had **10% losses** from shading.

4kWac × 5PSH × 0.9 derating = 18kWh/day

PSH = Peak Sun Hours = Insolation is the amount of sunshine that you get in a particular location on the average day throughout the year at a particular tilt angle. PSH is usually between 4 and 6.

0.9 derating is from the 10% losses.

If you lose 10%, then you keep 90%.

Mathematically, 10%/100%=0.1 (turn percentage into a decimal)

Subtract that from 1.

> 1−0.1=0.9 derating factor.
>
> Always think of what you are keeping with derating, not what you are losing.

Practice test question:

A 7kWdc PV system makes 5.5kWac when 1000W per square meter are shining on it. If this is in a location that will get an average of 4.5 peak sun hours (PSH) per day and there are 12% losses from shading, then about how much energy will this system produce in a month?

a. 145GWh
b. 650kWh
c. 830kWh
d. 90kWh

> Derating for 12% means you keep 88%
>
> 1−0.12 = 0.88, which is decimal for you keep 88%

Answer:

5.5kWac × 4.5PSH × 0.88 derating × 30days

Explanation:

System makes 5.5kWac (inverter output) under "peak sun conditions" of 1000W per square meter.

There are 4.5 peak sun hours of insolation, which means that the solar radiation is equivalent to 4.5 hours of 1000 watts per square meter. This is like if it were noon for 4.5 hours and then dark the rest of the day.

12% losses means that we lose 12% and we keep 88% of our production. Losses + what we keep has to equal 100%. Percentages are decimals moved 2 places. 12%/100=0.12 and is the same as 12%. If we subtract 0.12 from 1 we get what we keep, which is 0.88 or 88%. One way to check our math here is adding 0.12 + 0.88 and we get 1.

We then multiply:

5.5kWac × 4.5PSH × 0.88 derating × 30 days = 653kWh

The answer in an exam can be rounded off to 650kWh.

SYSTEM LOSSES

There are always going to be losses and when you know how much solar energy lands on the PV array, then every step of the way, there are going to be losses. If you have a 10% efficient solar system powering a 10% efficient lightbulb, you would end up with 1% of the energy from the sunlight being turned back into light. This is an example that is not too far from reality.

Here is the math:

10% in decimal form is 0.1

$0.1 \times 0.1 = 0.01 = 1\%$

Using sunlight for lighting is about 100 times more efficient than using PV and light bulbs.

There is a website put out by the US government National Renewable Energy Labs (NREL) that is often used to size PV systems. Here at the PVWATTS version 1 website we can see a table of various **derating factors**.

NREL PVWATTS websites: http://pvwatts.nrel.gov/

http://rredc.nrel.gov/solar/calculators/pvwatts/version1/

All of the derating factors can be multiplied together to come up with a single derating factor.

$0.95 \times 0.92 \times 0.98 \times 0.995 \times 0.98 \times 0.99 \times 0.95 \times 0.98 \times 1.0 \times 1.0 = \mathbf{0.77}$

When all of the derating factors above are multiplied together, the system derate value will be **0.77**, which means that the PV system will keep 77 percent of the energy or in other words lose 23% of the energy.

Figure 7.1 PV system derating factors from National Renewable Energy Lab (NREL)

Component Derate Factors	Default Value	Allowed Range
PV module nameplate rating	0.95	0.800–1.050
Inverter and Transformer	0.92	0.88–0.98
Module mismatch	0.98	0.970–0.995
Diodes and connections	0.995	0.990–0.997
DC wiring losses	0.98	0.970–0.990
AC wiring losses	0.99	0.980–0.993
Array wiring losses	0.95	0.300–0.995
System availability	0.98	0.000–0.995
Array soiling	1.000	0.000–1.000
Tracker misalignment	1.000	0.950–1.000

Unlike the previous example, these derating factors are from the dc STC wattage measurements of the PV modules to the ac kWh output. Every step along the way, you lose energy.

AC kWh output is what you want in the end and dc STC watts are how the PV modules are tested, rated and sold.

In the example above, the derating factor default value for shading is 1, which means that there is no shading. In most installations there is some shade. Shading will usually be calculated with a shade measuring device, such as a Solmetric Suneye or a Solar Pathfinder or with calculations and drawings.

Other **parameters that effect production are tilt and azimuth, which are collectively called orientation**. Orientation's effect on production is most accurately calculated with software. There are however generalities or "rules of thumb" that are often used.

TILT

Latitude tilt is best for equinox time of the year (around March or September 21). This is when the sun is directly over the equator. Also latitude tilt is good for annual production.

Latitude +15 degrees tilt is best for winter production. Often stand-alone PV systems are optimized for winter. If December were the month with worst insolation (solar energy), then we would call December the **"critical design month"**.

Latitude -15 degrees tilt is best for summer. This is when the sun is highest in the sky and decreasing the tilt is best.

A tilt that is close to zero can have trouble keeping clean, since the rain has trouble getting the dirt over the frame edge. Some designers do not tilt less than five or ten degrees, so that PV modules will self-clean better in the rain. In locations without rain for a whole season, dirt can build up on the module, which we call **soiling**.

Often designers will design a system so that the tilt is less than optimal, so that they can fit more on the roof. **An increased tilt angle takes more room, due to inter-row shading**.

AZIMUTH (NORTH, SOUTH, EAST AND WEST)

In the northern hemisphere, we tend to get the most production with PV facing **south**. Often other than south orientations are acceptable, depending on the orientation of the roof and the price of the energy. If PV prices are low and energy prices are high, such as in Hawaii, there is a good argument for filling up every space on the roof.

A reason to orient PV to the southwest is because of time-of-use (TOU) electricity rates. TOU rates make electricity more expensive in the afternoons when electricity demand is high and when the sun is moving to the west.

Here is a list of factors that would increase or decrease PV system output.

Increase output:

1. Cold temperatures (increases voltage)
   a. Increased airflow
   b. Space between PV and roof
   c. Cold climate
   d. High elevation

2. Clean PV (increases current)
   a. Rain recently
   b. Washing PV

3. Increased irradiance (increases current)
   a. Reflections
   b. High elevations

4. Efficient equipment
   a. Inverters
   b. Transformers
   c. Charge controllers

Decrease Output:

1. Soiling (decreases current)
   a. Dusty roads
   b. Farm animals

2. Snow on PV (decreases current)

3. Hot temperatures (decreases voltage)
   a. Lack of airflow
   b. BIPV (Building Integrated PV)

4. Voltage drop (wires too thin or long)
   a. Wire sizes: 2/0AWG > 1AWG > 10AWG

5. Loose wire connections

6. Shading

7. Inverter/equipment efficiency (inefficiency)

Practice question 1:

A 12kWdc PV system with 15% losses in a location with 4.8 Peak sun hours per day will make how much energy in a year?

12kWdc × 0.85 derating × 4.8 PSH × 365 = 17,870kWh per year

Practice question 2:

A PV system that makes 5kWac during peak sun conditions is in a location with 4.8 peak sun hours per day. How much energy will the system make in a month?

5kWac × 4.8PSH × 30 = 720kWh per month

Practice question 3:

A barn has a 100W light bulb that is on 100% of the time and a **746-watt (1HP)** pump that is on 25% of the time. How much energy does it use in a week?

Light

100W × 24 hours × 7 days = 16,800Wh = 16.8kWh

> **746W = 1 Horsepower**

Pump

746W × 24 hours × 0.25 × 7 days = 31,332Wh = 31.3kWh

Light + Pump

16.8kWh + 31.3kWh = 48.1kWh

Practice question 4:

If a 12V pump drew 5A for 2 hours, how much energy did it use and how many amp hours would it take from a 12V battery?

Energy calculation

12V × 5A × 2 hours = 120Wh (not kWh)

There are **different ways to calculate Ah** for this problem.

120Wh/12V = 10Ah
5A × 2 hours = 10Ah

Remember that Ah are just Amps × hours

Practice question 5:

Grid tied inverter size is based on?

a. Loads

b. PV array size in kW
c. PV array size in kWh
d. Charge controller

Explanation: A grid tied inverter will convert dc to ac at the rate of which the array is supplying power to the inverter. The inverter should be sized based on the array. In most instances, the inverter will be sized slightly smaller than the array. This is because the array is rated for dc STC watts and the inverter is rated for ac watts. **The grid tied inverter is sized based on the array size in kW.** Correct answer is b.

Practice question 6:

What would be the main benefit of oversizing a battery bank?

a. Increase in array kWh
b. Increase in days of autonomy
c. Decrease in array size
d. Decreased inverter size

Explanation: Increasing the size of the battery bank allows for more time with bad weather. Increasing your days of autonomy means that you can go more days without input from the sun. Autonomy means alone. **The main benefit of oversizing a battery bank is increasing days of autonomy.** Correct answer is b.

Practice question 7:

When sizing a **self-regulating PV system**, what criteria do you need to have in order to not need a charge controller?

a. Battery must be lithium ion
b. Battery bank must be less than array size
c. 1 hour charge is less than 3% of battery capacity
d. Battery must be Valve Regulated Lead Acid

Explanation: A self-regulating system does not have a charge controller and must not overcharge the battery. If a system is sized so that in **1 hour the battery will not get charged more than 3%** then this would be good for the battery. This is a rule spelled out in the National Electric Code. The correct answer is c.

Practice question 8:

Which 12kW system would produce the most energy?

a. BIPV
b. Flush mounted rooftop
c. Ground mount
d. Pole mount

Explanation: BIPV (Building Integrated PV) will have the least amount of air flow and operate the warmest and least efficient of the PV listed above. Then flush mounted rooftop, which is the typical residential solution, would be next on the list. Ground mount would produce even more and a **pole mount would operate best** of the above, because it would have the best airflow to release the heat away from the dark colored PV modules, so that they will be cooler and have more voltage. The correct answer is d.

Practice question 9:

How much 15% efficient PV will fit on 25 square meters of roof space?

a. 3.75MW
b. 4kW
c. 3750W
d. 55kW

Explanation: The average PV module installed is about 15% efficient. When we test PV, the input is 1000W/m^2. If the module is **15% efficient**, we would expect to get:

$$0.15 \times 1000W/m^2 = 150W/m^2$$
$$150W/m^2 \times 25\ m^2 = 3750W$$

The correct answer is c.

Practice question 10:

What is the simplest type of PV system?

a. Utility interactive
b. Stand-alone

c. Direct coupled
d. Bimodal

Explanation: Of the above answers, the most difficult is the bimodal. Bimodal means that it can work in utility interactive mode and stand-alone mode. Next on the list is stand-alone, since it is complicated to incorporate batteries. Utility interactive is next, since it has no battery or charger, but it does have an inverter. The simplest type of system is the direct-coupled system, which is just a load and a PV. It works when the sun is shining and does not when it is dark. A good example of this simple direct-coupled system is a solar attic fan.

PV system electrical design

SINGLE LINE DIAGRAMS

Single line diagrams (SLD) also known as one line diagrams are diagrams showing how components are connected to each other. On the other hand three line diagrams show more precisely how equipment is wired together. With every circuit, there are at least two and often three or more wires connecting the circuit and the grounding systems.

There are many different kinds of single and three line diagrams. Many three line diagrams have more than three lines. Some single line diagrams are very simple as the one shown in Figure 8.1 and others are more complex with a lot of equipment.

Figure 8.1 Single line diagram

Figure 8.2 Three line diagram

tem electrical design

Figure 8.3 Single line diagram fill in forms from www.solarabcs.org. The Solar America Board of Codes and Standards has an expedited permit process that is very helpful. Notice this drawing does not have separate lines for negative and positive.

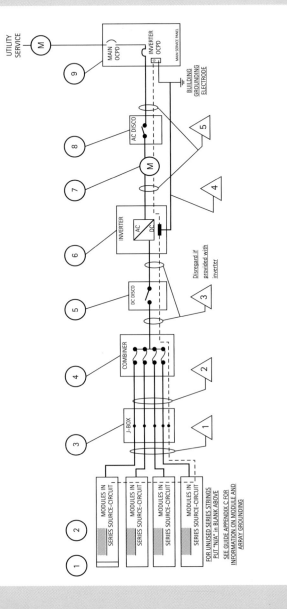

To control undercharging of batteries, the power has to turn off when the voltage of the battery gets low. This **can be done by the inverter or the charge controller**.

Figure 8.4 Single line drawing of typical stand-alone setup

Whatever the **low voltage set point is will determine the wire size for the inverter input circuit (between the inverter and the battery).** An inverter will require more current at a lower inverter voltage set point since:

Voltage × Current = Power

Just remember, the lower the voltage, the higher the current and the bigger the wire required for an off-grid inverter input circuit.

Figure 8.5 Stand-alone inverter input circuit current determined by low voltage

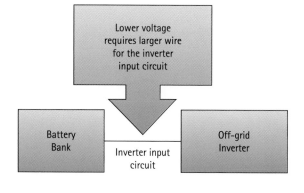

Figure 8.6 Series positive to negative increases voltage

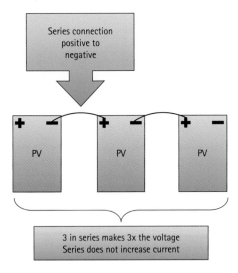

Figure 8.7 Parallel connections increase current not voltage

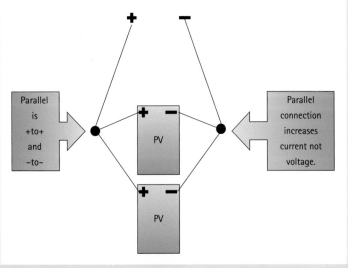

PV source circuits meet **PV output circuits** at the **combiner box**. This is where the parallel PV circuit connections are made.

AMPACITY

The **ability** of a conductor (wire) to **carry current** is called **ampacity**.

When a conductor carries more current, it heats up. The insulation (like plastic) around the wire can only take so much heat. When a conductor carries too much current, it heats up beyond the temperature rating of the insulation of the conductor. The **temperature rating of the conductor affects ampacity**.

Factors that influence ampacity (ability of conductor to carry current):

1. Thickness of wire
2. Insulation around the wire
3. Hot temperatures
4. Airflow around wire.

OHM'S LAW AND VOLTAGE DROP

Ohm's law states **V = IR**

V = Voltage
I = Current
R = Resistance

Voltage drop is when voltage and power are lost on a wire.

Voltage drop is reduced when the wire is either short or thick, which is why it is better to use thicker wires for better efficiency.

Because V = IR, if we have less current, then we will have less voltage drop. This means that **voltage and current are directly proportional** to each other according to Ohm's law.

It should be obvious that having a short wire is good and having a thicker wire is more efficient.

Also, having higher voltage is more efficient, since Voltage × Current = Power and more voltage means less current for the same amount of power. This is why we have high voltage power lines.

Transformers change high voltage and low current to low voltage and high current or vice versa. Transformers only work for alternating current (ac).

DC to dc converters can do the same thing as a transformer for direct current.

Some inverters have transformers and some do not. An inverter without a transformer can be called a few things.

Inverters without a transformer are called:

1. **Transformerless inverters**
2. **Ungrounded inverters**
3. **Non-isolated inverters.**

Ungrounded inverters must be used with PV wire for exposed wiring and USE-2 wire cannot be used exposed. (USE-2 wire is often used for inverters with transformers.)

Ungrounded inverters have no grounded conductor (white wire) on the dc side of the inverter.

Ungrounded inverters do have equipment grounding. Equipment grounding is a green wire or a bare copper wire.

Table 8.1 Ungrounded and grounded inverters

Ungrounded Inverters	Grounded Inverters
Cannot use USE-2 exposed	Can use USE-2 exposed
Must use PV wire when not in conduit*	Can use PV wire exposed
No dc white wire	Must have dc white wire
Has equipment grounding	Has equipment grounding

Note: *conduit is pipe that wire is put into in order to protect it. Also conduit is a raceway.

The reason that an ungrounded inverter is called ungrounded, when it does have equipment grounding is very confusing. Technically an ungrounded inverter does not have dc "system" grounding, which means it **has no dc white-grounded conductor**.

CHARGE CONTROLLERS

Only used with battery systems (something to charge).

Are between the PV and the battery.

When a charge controller is used an inverter is usually not connected to the PV.

Charge controllers are **not** used on **self-regulating PV systems** where charging current in 1 hour is **less than 3%** of battery capacity.

AC AND DC COUPLED BATTERY SYSTEMS

DC coupled systems are the typical battery based PV systems where PV is connected to a charge controller that charges a battery.

AC coupled systems require two different kinds of inverters. They use utility interactive inverters that can be turned on by another battery-based inverter. In a way, the battery-based inverter tricks the utility interactive inverter into thinking that there is a utility present in order to turn on the utility interactive inverter. AC coupled systems can be made to convert a utility interactive system into a system that works on- and off-grid, which is also called a grid tied battery backup system.

AC and dc coupled systems can both be configured to work as grid tied battery backup systems. Another name for this is a **multimodal** or **bimodal** system. This means it can work in utility interactive or stand-alone mode. When there is a utility outage, the multimodal inverter will power only loads that are isolated from the utility.

BATTERY CHARGING STAGES

Bulk is charging fast and done first.

Absorption charge is slowing down the charge.

Float charge is a trickle charge to keep the battery charged.

Equalization charging is for battery maintenance and is the **highest voltage** charge.

TYPICAL USA UTILITY VOLTAGES

Residential 120/240V split phase/single phase

Commercial 3-phase Voltages:

> 120/208V
> 277/480V
> 240V
> 480V

If you ask an honest electrician how long it takes to learn about 3-phase power, they will tell you they are still learning, so do not get discouraged.

OVERCURRENT PROTECTION DEVICES (OCPD)

Two types of OCPD are fuses and circuit breakers.

OCPDs protect conductors and equipment from overcurrents by opening the circuit when the current is too much. (Opening a circuit is how things are turned off.)

Maximum Series Fuse Rating is what is specified on the back of a PV module (along with rated power, open circuit voltage, maximum power voltage, short circuit current and maximum power current). The maximum series fuse rating means that if you had a fuse with a higher current, it would not protect the PV module. In a way, the maximum series fuse rating is like the "ampacity" of the PV module.

Module interconnects connect the modules together and can be made with

USE-2 wire or **PV wire**. They are also protected by the fuse mentioned above. USE-2 cannot be used for "ungrounded" systems.

VOLTAGE RANGES

Equipment that PV is connected to and process power, such as **inverters and charge controllers**, will have a voltage range for operation. If the voltage is too low, then the equipment will not turn on. If the voltage is too high, then it can damage the equipment or be dangerous.

The voltage ranges are specified for the equipment on labels and datasheets.

VOLTAGE AND TEMPERATURE CALCULATIONS

The most difficult concept to master in this book is what is commonly called "string sizing". What we will do is figure out what the voltage will be when the voltage goes up when it is cold or what the voltage will be when it is hot.

When it is **hot and the voltage gets low**, we run the risk of the system not turning on or producing, which is not a safety problem, but a waste.

When it gets too **cold, the voltage will get higher** and it is possible that on a cold day the PV system will have enough voltage to destroy equipment, void the warranty of the inverter, cause a fire or violate the National Electric Code. Overvoltage is a safety issue.

There are different ways of determining how voltage changes with respect to temperature. The main way that this is done is with what is called a temperature coefficient for voltage.

Coefficients are also used for other things in science and with the electricity.

EXAMPLES OF COEFFICIENTS

Expansion of metal: Metal expands a certain amount for each degree warmer it gets. We can use the coefficients for the expansion of aluminum and steel when designing our PV racks, rails and conduit.

Atmospheric pressure: For every meter rise in elevation, you have less atmosphere. This can be important when designing a high altitude PV system.

We can even have coefficients for making coffee. For every six cups of water, we use one scoop of ground coffee. If we want 24 cups of coffee, then we need three scoops. We could say the coefficient is one sixth of a scoop per cup of coffee or 0.167 scoops per cup (forgive me if you like it stronger).

PV AND VOLTAGE

Now let's think of PV and voltage. In general PV open circuit **voltage (Voc) goes up about one third of a percent for every 1°C** decrease in temperature.

Since all of the PV is tested at **STC at 25°C** then at 24°C we would gain one third of a percent of our voltage. One third of a percent is represented as 0.33%/°C. If our voltage was 100V then one third of a percent of 100V is one third of 1, since one percent of 100 is 1.

One third of 1 is 0.33, so at 24° our 100V would go up to 100.33V.

If our temperature went down ten times that much or ten degrees less than 25°C to 15°C, then ten times a third is ten thirds or 10 × 0.33 = 3.3% increase in Voc.

3.3% of 100V is 3.3V, so then our new voltage would be 103.3Voc.

If we really thought about it and our temperature coefficient for voltage is a third of a percent for each 1°C, then for every 3°C is a 1% change.

What we are going to do now is a practice problem. This will seem difficult the first time so do not get discouraged. Once you get the system down, you may think it is easy.

Given:

PV Voc = 50V
Temp Coefficient for Voc = -0.33%/°C
Cold Temperature = 0°C

What will be the voltage?

Change in temperature from 25°C to 0°C is 25 less or -25°C.

Multiply change in temperature by temperature coefficient:

–25°C × –0.33%/°C = 8.25% increase in Voc

An 8.25% increase in voltage can be found by multiplying the voltage by 0.0825.

We get 0.0825 by moving the decimal 2 places to the left or by dividing 8.25% by 100.

0.0825 × 50V = 4.125V increase in Voc

We can then add 4.125V to 50V to get our voltage

50V + 4.125V = 54.1V (rounded off)

A short cut to doing this is to multiply our voltage by 1.0825 instead of 0.825 to get our final voltage.

50V × 1.0825 = 54.1V

54.1V is our open circuit voltage (Voc) at our cold temperature of 0°C.

The next question is how many of these PV modules can we put in series. We first need to know the maximum acceptable voltage.

Inverter maximum voltage is 500V

To determine how many modules can be in series without exceeding 500V

500V / 54.1V = 9.24 modules

Since we cannot cut a module in half or go over voltage, then we will always round down when determining maximum number of modules in series.

If you are using high temperatures and determining minimum number of modules in series, you will round up.

The maximum number of modules in series in this example is 9 in series. 10 in series would bring the voltage up over 500V on a cold day and void the warranty of the inverter or worse.

At the end of this book, there will be a special section on doing voltage temperature calculations where we will study this concept more.

PV system mechanical design

PV Mounting Systems affect performance due to:

1. Airflow
   a. More airflow is better and will cause heat to leave PV and make voltage higher. High voltage means more power and energy because:
      i. **Volts × Amps = Power**
      ii. **Power × Time = Energy**

2. Orientation
   a. Residential roof mounts usually go with the slope of the roof. **When a system goes with the slope of the roof, we call that flush mounted.** In most parts of the USA a **30 degree sloped PV system facing south is close to optimal**.
   b. Repeating rows take inter-row shading into consideration. When the tilt is less, more PV will fit because of less inter-row shading with decreased tilt and module height.
   c. Flat or very low-sloped PV systems will have trouble staying clean and will have soiling or will have trouble shedding snow.

Building Integrated PV (BIPV) is PV that is part of the roofing or building materials. BIPV does not have airflow, so will operate at a higher temperature, which will decrease performance when compared to systems that have airflow. BIPV is more expensive and less common than mass-produced modules.

Flush Mounted PV usually has three inches to six inches between the PV and the roof. The six inches keeps the PV cooler than BIPV and will also keep the roof and the attic cooler than if the sunlight was hitting the roof directly. Flush mounted PV is the predominant residential PV mounting system.

Figure 9.1 BIPV
Source: ©2014 Tony Diaz, Century Roof and Solar

Flush mounted PV mounting systems:

1. Composition asphalt shingle mounting
2. Tile mounting
3. Metal roof mounting
4. Wood shake shingles.

Waterproofing is always a concern when putting holes into a roof and there are many different types of systems used to waterproof roof penetrations.

Flashing is strongly recommended.

Low slope roof mounting is the typical **commercial roof** mounting system. A low slope roof is often called a flat roof, although it is never completely flat so that rainwater will drain from the roof.

There are two predominant types of low slope roof mounting systems:

1. **Ballasted systems** are **not structurally attached** to the roof with penetrations (holes). Ballasted systems are designed so that something heavy is put onto the PV system to hold it down. Often concrete rectangular paving stones are used. Ballasted systems are often not used when the roof cannot take the extra weight or where there is a lot of wind.

Figure 9.2 Flush mounted composition asphalt shingle PV mounting system with flashing. Flashing is overlapping of materials to prevent water intrusion. Figure shows QuickMountPV flashing with IronRidge rail system.
Source: Courtesy Quick Mount PV

Figure 9.3 Ballasted system
Source: Courtesy Advanced Solar Products, Solstice Mounting System.

2. **Penetrating Systems** use structural attachments, usually called posts, jacks or pedestals that attach the racking system to a strong part of the roof. A strong part of the roof is usually a purlin, a truss or a rafter, which can take the forces of the PV system. It is not good to attach the PV system to something not as strong, such as the roof deck. Most residential systems are also penetrating systems and attach to the roof via structural attachments.

3. **Combination Ballast and Penetrations**. Often engineers will put as much weight as the roof can handle and then put some penetrations around the corners where the most wind uplift forces are. (Wind uplift forces are measured in pounds per square foot.)

Ground mount PV systems contain most of the PV installed in the world today in large "Utility Scale" solar farms. Ground mounts take up valuable real estate and are not as common in cities. Ground mounts come in all sizes from less than a kW to hundreds of MWs.

Ground mounted systems can be mass-produced with increased efficiencies.

Ground mounts have increased airflow beyond roof mounted systems, which means greater voltage, power and energy.

Pole mounted PV systems are PV arrays that are put up on a single pole. They have **more airflow than ground mount, and roof mounted systems** but are more difficult to engineer. Pole mounted systems can more easily fall over if not secured properly.

Tracking systems follow the path of the sun for increased production. As the price of PV goes down, the need for tracking systems decreases. When PV was expensive every extra kWh of energy that we could get out of a module was important. Now that PV is less expensive, it is often a better value proposition to buy more PV than to buy a tracker. Trackers are also known for being high maintenance.

All of that being said, the most popular tracker used today is a **single-axis tracker** that will face east in the morning, west in the evening and will be flat at noon. This is called a **horizontal axis tracker**.

A **2-axis tracker** will be mounted on a pole mount and **follow the sun in every direction exactly**.

Concentrating PV is PV that uses magnifying lenses or mirrors to focus light onto an efficient solar cell. Concentrating PV **must** have a **2-axis tracker** to accurately track the sun in order for the concentrating PV to work well.

Sealants are used to keep the roof waterproof. The most popular sealants used are **polyurethane and silicone**.

Fasteners are used to fasten the mechanical racking systems. **Stainless steel fasteners** are usually used and most importantly used in **humid, moist or marine environments** where corrosion is more likely.

Different forces are considered when mounting PV systems. Some of these forces are wind loads, snow loads, seismic loads, live loads and dead loads.

Wind loads are measured in **pounds per square foot (Psf)** in the USA. Wind loads are worse in different parts of the roof. Typically the corners are worse, the edges are next and the middle of the roof is the safest place for PV. Also, urban areas are safer than wide open areas due to wind. Increased module tilt causes more forces on the PV from the wind.

Snow loads add extra weight. Roofs in these areas are often not able to take extra weight from ballasted PV systems. **Structural engineers** often have to qualify a roof for a PV system in areas with high snow loads. Lower tilt angles are often used in areas with snow loads, since high tilt angles can cause build up of snow on the roof, which can be too much weight for the roof.

Dead loads are loads that are **permanent** on the roof, such as the solar system you install and anything that is part of the building.

Live loads are **temporary** and include the workers on the roof and the snow falling on the roof.

Chapter 10

Performance analysis, maintenance and troubleshooting

If we have a good grasp of the previous nine chapters, this chapter will be common sense.

Many PV systems were installed incorrectly and many others were installed in ways that will not last the test of time. Due to the rapid growth of PV installations, most of the equipment installed in the world is less than two years old. As equipment ages, it will have problems that need fixing.

Commissioning a PV system is what is done after the system is installed. There are different levels of commissioning. Just turning the system on after checking voltage is the simplest form of commissioning, but usually when someone talks about commissioning a system, especially a large system, they measure irradiance data in the **plane of the array (tilt and azimuth orientation)** with an irradiance-measuring device called a **pyranometer**. Also, they measure the solar cell temperature. Then calculations are done to correct for temperature and irradiance, in order to see that the PV system is working as it should. A common mistake people make is assuming that a 7kW PV system with a 6kW inverter should be making 6 or 7kWac when the system is powered up. More often a PV system will be making half of its rated power rather than its full rated power. This is because clouds, time of day and various derating factors, including solar cell temperature, irradiance, soiling, inverter inefficiency and voltage drop will reduce the amount of energy produced.

Another common reason that a PV system is not turning on is because the system needs to be turned on for five minutes before it will export power to the grid. This is because of the anti-islanding requirements of utility interactive inverters.

Monitoring is very common with PV systems. Often residential monitoring consists of a device that measures current and voltage on the ac output of the inverter and relays that information via the Internet.

Monitoring for larger systems often has more parameters that are measured, such as wind speed and irradiance. Some monitoring is on the source circuit level and other monitoring would be on the inverter output level.

When diagnosing a problem on a PV system, make sure to **stay safe**. This means looking at your system first, turning things off with disconnects and then taking measurements and saving the corrections for last when you know what is wrong. Always remember to be safe! Making sure that your voltmeter works before testing with it is standard practice.

Common performance issues:

1. PV system turns off on hot sunny days.
   a. Problem is voltage gets too low when temperature gets too hot.
   b. Correction is putting more modules in series to raise voltage or getting another inverter with a lower voltage input window.

2. Ground fault detected.
   a. Likely a fault on a source circuit.
   b. Measure voltages on separate strings to find ground fault.
   c. Make sure everything has been turned off first.
   d. Be careful of getting shocked.

3. Blown fuse in a combiner.
   a. Likely positive to negative fault.
   b. Multiple strings feeding backwards through one fuse in a combiner.
   c. Carefully correct the problem after visually determining what is wrong and turning everything off.
   d. Find the fault and correct the problem being careful to not get shocked.

4. Performance is not good for a few days in the winter according to monitoring.
   a. Look for snow or bad weather.
   b. Solution is to wait or move snow.

5. In a dry climate performance is worse at the middle or end of the dry season.
   a. Look for soiling (dirt).
   b. Solution is to wash modules.
   c. Check for low operating voltage problems due to heat on array.
   d. Check for excessive voltage due to undersized conductors.

Battery maintenance:

1. Check voltage with volt meter.
2. Clean terminals and equipment.
3. Check connections.
4. Inspect fuses and disconnects.
5. Inspect ventilation system.

Battery maintenance for flooded lead acid batteries that are not sealed:

1. Check fluid levels.
2. Replace missing fluid with distilled water.
3. Check specific gravity of battery acid.
   a. **Hydrometer** checks battery acid with a density test
   b. **Refractometer** checks battery acid with a test to see how light bends in fluid.
4. **Equalization charge** is a controlled overcharge for maintenance of non-sealed flooded lead acid batteries. Equalization charge splits H_2O and releases bubbles of hydrogen and oxygen, which will stir up the battery.

TEST TAKING STRATEGY

When taking a NABCEP Exam, every question has four possible answers: a, b, c or d.

- Use common sense. Many times you can rule out most answers without even doing the calculation. The wrong answers will have the wrong units, be in the opposite direction or obviously be way off.
- No more than one answer can be correct, so if there are two answers that mean the same thing, then probably they are both wrong.
- There will be no "all of the above" choices.
- Write notes onto the exam, so you can go back at the end and see if you have a different perspective. Many questions have answers to other problems in them.

- Remember that complex math is a series of simple problems.
- Math can be done in different ways.
- You should have plenty of time, so be patient and do not let a tough question destroy your confidence. You do not have to get everything correct to pass.
- For math problems, practice with real equations: do not expect to read a book and then be able to work equations without practicing.
- Take the exam well rested and if you consume caffeine on a regular basis, try not to have too much or too little.
- State dependent memory: study at the same time of day and conditions that you take your exam.
- If you cannot do voltage temperature correction math, you can still pass the exam.
- Have confidence!

Special section: voltage temperature calculations

STRING SIZING: KNOWING HOW MANY MODULES TO PUT IN SERIES

The most difficult part about this book will be learning to do the voltage temperature correction math. If you are overwhelmed with math, you can review this section and come back to it later. It is best to practice working sample questions in order to learn these concepts. Remember, these problems are a few simple calculations done in a pattern. The more you practice, the more confidence you will have.

For crystalline PV, Voc goes up about a third of a percent for every decrease of 1°C in temperature under 25°C.

> A change of 1°C is a change of 1.8F
> Cs are bigger than Fs

That means if it gets to be 3°C colder than 25°C then the change in temperature is going to be −3°C when it is 22°C.

A −3°C change with a third of a percent change in voltage for each degree C would mean:

3°C × one third percent per degree = 1% change

Since it is getting colder, that percentage will be a 1% increase in voltage.

6 degrees colder is a 2% increase in Voc

9 degrees colder is a 3% increase in Voc

12 degrees colder is a 4% increase in Voc

24 degrees colder than 25°C is 1°C

24 degrees colder is an 8% increase in Voc

These changes above are for a one-third percent coefficient. Sometimes the coefficients are different, but at least these examples are in the same ballpark. On an exam, if you are in the right ballpark, you have a good chance of answering a multiple-choice question correctly. In the field or on your desk at work, knowing the right range of answers is very helpful.

Here is another way of thinking of coefficients:

Say that all you had were t-shirts and for each 3 degrees below 25°C you had to put on another T-shirt to keep warm. How many shirts would you have to put on to keep warm at −5°C?

Answer:

● −5°C is 30 degrees colder than 25°C
● 30°C / 3°C per shirt = 10 shirts

When we design our systems, we have to determine how many modules we can connect in series, so as to not go over voltage. Going over voltage can break whatever your PV is connected to. Inverters, charge controllers, wires, disconnects and all electrical equipment have voltage limits, which we can not exceed safely.

If we put too many PV modules in series, then on a cold day when voltage is higher, it can ruin an inverter. We always have to design our systems so that we do not have too many PV modules in series. We call this string sizing.

The information that we use for string sizing is:

1. **Cold expected temperature**
2. **Temperature coefficient Voc**
3. **Voc of PV module**
4. **Maximum input voltage of inverter**

Coldest expected temperature many people get from www.solarabcs.org.

Temperature coefficient of Voc is read from the back of the PV module, from the datasheet/cutsheet. This is usually close to a third of a percent per degree C or −0.33%/°C.

Voc of PV module is found on the back of the module, the datasheet or many other places.

Maximum inverter input voltage is found on inverter documentation, cutsheet or label on the inverter.

Here is how to do the math!

Steps:

1. **Find Delta T** (delta means difference)
   a. Cold temp minus 25°C

2. **Multiply Delta T × Temp coefficient Voc**
   a. You then get % increase in voltage

3. **Determine increase in voltage**
   a. Think of this like sales tax
      i. 10% increase multiple by 1.1
      ii. 8% increase multiple by 1.08

4. **Max input voltage/cold temp PV voltage**
   a. Determines modules in series
   b. Always round down for cold temperatures, since rounding up would give you too much voltage.

Sample problem:

Given:

1. Cold temperature   −15°C
2. Temperature Coef Voc   −0.33%/°C
3. Voc   35Voc
4. Inverter Max input voltage   600V

How many is the most modules that can be configured in series?

−15°C −25°C = −40°C (Delta T)

−40°C × −0.33%/°C = 13.2% increase Voc

1.132 × 35Voc = 39.62Voc cold

600V/39.62V = 15.14

15 in series max

The best way to learn this is by repeating it many times right now. Here are some data and answers are in the back of the book. Solve for max modules in series.

Question 1:

1. Cold temperature — −10°C
2. Temperature Coef Voc — −0.33%/°C
3. Voc — 35Voc
4. Inverter Max input voltage — 600V

Question 2:

1. Cold temperature — −10°C
2. Temperature Coef Voc — −0.34%/°C
3. Voc — 35Voc
4. Inverter Max input voltage — 600V

Question 3:

1. Cold temperature — −10°C
2. Temperature Coef Voc — −0.34%/°C
3. Voc — 37Voc
4. Inverter Max input voltage — 600V

Question 4:

1. Cold temperature — −10°C
2. Temperature Coef Voc — −0.34%/°C
3. Voc — 37Voc
4. Inverter Max input voltage — 500V

Question 5:
1. Cold temperature   −5°C
2. Temperature Coef Voc   −0.35%/°C
3. Voc   22Voc
4. Inverter Max input voltage   550V

Question 6:
1. Cold temperature   −40°C
2. Temperature Coef Voc   −0.33%/°C
3. Voc   36Voc
4. Inverter Max input voltage   1000V

How to do this on calculator **without paper**
1. Cold temperature   −5°C
2. Temperature Coef Voc   −0.33%/°C
3. Voc   36V
4. Inverter Max input voltage   600V

A. −20°C−5°C = −30°C
B. −30°C × −.33%/°C = 9.9%
C. 9.9%/100% = .099 correction factor
D. 099 + 1 = 1.099
E. 1.099 × 36V = 39.564V cold temp
F. 39.564 and 1/X button = 0.0252755 (numerator becomes denominator)
G. 0.0252755 × 600 = 15.17 in series
H. 15 in series

String sizing can also be done for hot temperatures. String sizing for hot temperatures determines the shortest string. If a string is too short, the inverter will not have enough voltage. This is not a safety issue, just not a good idea to have your inverter not turn on. Since it is not a safety issue, it is not a code issue. When doing the math for a hot temperature you would use the Vmp, because that is the low voltage and we use a temperature coefficient for Vmp. Also we are going to find the shortest string and round up rather than down. Additionally, the PV often operates 20°C to 30°C hotter than ambient because it is sitting in the sun.

Example of calculation for hot temperature correction for low voltage:

1. Hot PV cell temperature   40°C
2. Temperature Coef Vmp   −0.4%/°C
3. Vmp   30Vmp
4. Inverter Min input voltage   250V

40°C −25°C = 15°C (Delta T)
15°C × −0.4%/°C = −6% (decrease) Vmp
1−0.06 = 0.94 is derating factor
(for 6% decrease)
0.94 × 30Vmp = 28.2Vmp hot
250V/28.2V = 8.7
9 in series min

70 practice exam questions

The NABCEP Entry Level Exam is a timed test, which you will be given 2 hours to complete.

1. An off-grid installation has a 1HP pump that is on 10% of the time and a 20W light that is on 100% of the time. What is the energy use per month?
   a. 54Wh
   b. 68kW
   c. 68kWh
   d. 0.63 MWh

2. What kind of ground faults do most inverters detect?
   a. High voltage
   b. Medium voltage
   c. AC
   d. DC

3. Current carrying ability of a conductor (wire) is called
   a. Ampacity
   b. Amps
   c. Inductance
   d. Insolation

4. According to ohms law and voltage drop, the relationship between voltage and current is
   a. Proportional
   b. Inversely proportional
   c. Directly inverse
   d. Inversely direct

5. An Array in Sacramento California in August is not producing as expected. Which of the following is NOT a likely problem?
   a. Decreased ac wire size and voltage loss
   b. String size too short
   c. String size too long
   d. Soiling

6. When using a digital multimeter, what is the first thing you should do?
   a. Turn the meter on
   b. Inspect the meter
   c. Touch the black lead to the white wire
   d. Touch the black lead to the black wire

7. Which is the worst design error?
   a. BIPV on a commercial job
   b. A fast growing tree to the north of the array in Idaho
   c. PV at 5 degree tilt at 40 degrees latitude
   d. Different tilt angles in a PV source circuit

8. The low design temperature is −20°C and the Voc temperature correction factor for your PV is −0.34%/°C. The characteristics of your module are Isc = 8A, Voc = 37V and Vmp = 29V. How many modules can you put in series for an inverter that cannot go over 500V?
   a. 10
   b. 11
   c. 12
   d. 13

9. MPPT is for
   a. Optimizing energy from the utility
   b. Getting the most energy from a battery with a MPPT inverter
   c. Optimizing power from a PV array
   d. Tracking the sun with a 2-axis tracker

10. Concentrated PV
    a. Only works with direct sunlight
    b. Usually has a single axis tracker
    c. Will get 30% power during cloudy conditions
    d. Is using heat to make electricity

11. Fall protection should be used at heights over
    a. 4 feet
    b. 2 feet
    c. 3 feet
    d. 6 feet

12. Parallel PV circuit connections increase
    a. Voltage
    b. Current
    c. Resistance
    d. EMF

13. Optimal tilt for off-grid PV system in the winter
    a. Latitude
    b. Latitude −15 degrees
    c. Latitude +15 degrees
    d. Vertical

14. For a transformerless inverter, which type of wire should be used when exposed to the elements outdoors on PV source circuits?
    a. PV wire
    b. THWN-2
    c. USE-2
    d. PV-2

15. Bypass diodes are located in the
    a. Combiner box
    b. Inverter
    c. Wiring harness
    d. PV module

16. A PV array at a higher elevation would tend to produce more
    a. Current
    b. Heat
    c. Resistance
    d. Polysilicon

17. Voc = 36V, low temp = −2°C, inverter maximum input voltage = 500V, temp coef voltage = −0.35%/°C. What is the maximum number of modules that can be connected in series?
    a. 13
    b. 10
    c. 11
    d. 12

18. How much 14% efficient PV fits on 25 square meters?
    a. 3000W
    b. 4.5kW
    c. 4000W
    d. 3.5kW

19. Which is the smallest conductor?
    a. 18 AWG
    b. 10 AWG
    c. 0 AWG
    d. 3/0 AWG

20. Which of the following is the simplest type of PV system?
    a. Utility interactive
    b. Stand-alone
    c. Self-regulating
    d. Direct coupled

21. Sun path charts differ based on
    a. Longitude
    b. Latitude
    c. Azimuth
    d. Tilt angle

22. Series connections increase
    a. Voltage
    b. Current
    c. Cold
    d. Heat

23. The longest shadow at 2PM would be on
    a. June 1
    b. November 12
    c. February 14
    d. March 29

24. What is the ground mount advantage over a rooftop system?
    a. Takes no extra real estate
    b. Cooler
    c. Distributed generation
    d. Soiling

25. Which test conditions have higher voltage?
    a. STC
    b. PTC
    c. CEC
    d. NOCT

26. Which is the best tilt angle for annual production in most of the USA
    a. Latitude +15 degrees
    b. Latitude −15 degrees
    c. 30 degrees
    d. Longitude tilt

27. A 100kW STC PV system with 8% system losses, using PV that is +3 −0% production tolerance, a 95% efficient inverter, what would be the expected ac output if 1000W/m² irradiance?
    a. 84.8kW
    b. 100kW
    c. 87.4kW
    d. 88.7kW

28. A 12V light is left on for 13 hours at 1.5A with a 12V battery. How much energy is used?
    a. 23 Ah
    b. 0.234 kWh
    c. 23.4 Wh
    d. 19.5 Ah

29. In the question above, how many Ah are used?
    a. 23 Ah
    b. 0.234 kWh
    c. 23.4 Wh
    d. 19.5 Ah

30. Electrical safety is best studied at
    a. OSHA website
    b. OSH website
    c. CSLB website
    d. UL website

31. The interconnection of a PV system is approved by
    a. City
    b. County
    c. State
    d. Utility

32. Grounding (bonding) of a transformerless PV system on a roof should be done to
    a. Stainless steel hardware
    b. Flashings
    c. Racking and aluminum frames
    d. Grounded conductor

33. Lowest grid voltage
    a. Transmission
    b. Distribution
    c. Service equipment
    d. Generation

34. If a 12V PV module has 36 cells, then how many volts is a 54-cell module?
    a. 6V
    b. 18V
    c. 24V
    d. 17V

35. Transformers convert
    a. High voltage low current to low voltage low current
    b. Low voltage low current to high voltage high current
    c. High voltage low current to low voltage high current
    d. Low current low current to high voltage low current

36. Undersized PV array
    a. Decreases voltage drop
    b. Decreases battery life
    c. Increases current drop
    d. Increases resistance

37. Bypass diodes
    a. Reduce current
    b. Limit effects of shading
    c. Are wired in series with groups of solar cells and prevent reverse current
    d. Are in the combiner box

38. Microinverters are connected between the _____ and the _____.
    a. PV module, service panel
    b. String, combiner
    c. Main breaker, backfeed breaker
    d. Utility meter, transformer

39. A system that sits on a flat/low slope rooftop and has no penetrations is called a
    a. Flush mounted system
    b. Flashed system
    c. BIPV system
    d. Ballasted system

40. What is the proper slope of an extension ladder?
    a. 1:3
    b. 1:4
    c. 1:5
    d. 1:6

41. Voc = 22V, Vmp = 18V, hot PV cell temp = 50°C, inverter operating voltage range = 250V to 600V, temp coefficient of voltage = −0.38%/°C. What is the least amount of modules that should be connected in series?
    a. 12
    b. 14
    c. 16
    d. 18

42. The primary power source for a utility interactive PV system is the
    a. Inverter
    b. Utility
    c. PV
    d. Charge controller

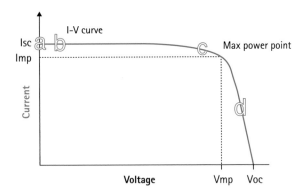

43. Which place on the IV curve above produces power and has the most current?
    a. A
    b. B
    c. C
    d. D

44. What is specified on a charge controller
    a. Irradiance limit
    b. Insolation limit
    c. Maximum voltage
    d. Minimum current

45. Units for irradiance are
    a. Energy per area
    b. Power per area
    c. Volts per square meter
    d. Amps per square meter

46. What is used to calculate maximum system voltage?
    a. Current
    b. Power
    c. Temperature
    d. Irradiance

47. What will be increased by reflective surfaces (albedo)?
    a. Voltage
    b. Resistance
    c. Current
    d. AM

For the following 3 questions, use the information from the table below:

Question 48 to 50 Table

Power	250W
Open Circuit Voltage	40V
Maximum Power Voltage	35V
Short Circuit Current	8A
Maximum Power Current	7.5A
Temp Coefficient for Voltage	−0.33%/°C
CEC Power	220W
NOCT	48°C
Inverter low MPPT voltage	230V
Inverter maximum input voltage	550V

48. There are 4 PV source circuits of 10 modules each. What is the MPPT voltage of the array at Standard Test Conditions?
    a. 160V
    b. 350V
    c. 140V
    d. 230V

49. What is the power of the array?
    a. 10kW
    b. 110kW
    c. 1000W
    d. 40kW

50. What is the maximum number of modules in series at a location with a low temperature of minus 40°C?
    a. 9
    b. 10
    c. 11
    d. 12

51. What is the best voltage for charging a 12V battery?
    a. 12V
    b. 11.8V
    c. 12.3V
    d. 14.1V

52. Unit for resistance?
    a. Amp
    b. Volt
    c. Om
    d. Ohm

53. Wind uplift forces in the construction industry in the US are measured in
    a. Pounds per square foot
    b. Miles per hour
    c. Square feet per pound
    d. kW per m^2

54. Direction on a compass is
    a. Latitude
    b. Longitude
    c. Meridian
    d. Azimuth

55. Hydraulic analogy for voltage is
    a. Pressure
    b. Flow
    c. Volume
    d. Drag

56. Which room cannot have a PV dc disconnect?
    a. Kitchen
    b. Attic
    c. Bathroom
    d. Bedroom

57. Stand-alone inverter output goes to
    a. Loads
    b. Utility
    c. Meter
    d. PV

58. Best sealants for residential rooftop
    a. Polyurethane and silicone
    b. Tar and paraffin
    c. Bitumen and rubber
    d. Silicon and caulk

59. 500W/m^2 for 4 hours is
    a. 20kWh
    b. 2kWh
    c. 2kW
    d. 5kW

60. For a supply side connection, the output of the inverter is limited by
    a. Service entrance conductors
    b. Load side breaker
    c. PV dimensions
    d. Combiner box ampacity

61. Which is the smallest installable part of a PV system?
    a. Solar panel
    b. Solar module
    c. Solar array
    d. PV source circuit

62. What is the typical hazard with PV systems?
    a. Shock
    b. Arc flash
    c. Lightning
    d. Heat stroke

63. A 4kW PV system with 15% losses in a location with 4.5 sun hours per day, what would be the annual output?
    a. 985.5kWh
    b. 1450kWh
    c. 5585kWh
    d. 4550kWh

64. If available roof area is 12m² and the PV is 15% efficient, then how much PV will fit?
    a. 4.5kW
    b. 2.2kW
    c. 1.8kW
    d. 3.3kW

65. Which of the following is a typical direct PV system application?
    a. Lighting
    b. Batteries
    c. Pumping water
    d. Fire alarm

66. Current is affected mostly by
    a. The number of solar cells in series
    b. The size of the solar cell
    c. The backsheet of the solar module
    d. The temperature of the cell

67. In California the north needle of the compass will point
    a. East of true north
    b. West of true north
    c. North of true north
    d. South of true north

68. Maintenance of sealed valve regulated lead acid batteries includes
    a. Checking fluid levels
    b. Equalization charge
    c. Scrubbing lead plates
    d. Cleaning terminals

69. Given that Voc is 18V and the temperature correction factor = 1.2. What would be the maximum system voltage if there were 16 modules in series?
    a. 340V
    b. 346V
    c. 240V
    d. 440V

70. STC watts can be calculated by
    a. Vmp × Voc
    b. Voc × Isc
    c. Vmp × Imp
    d. PTC × CEC

ANSWERS

1. An off-grid installation has a 1HP pump that is on 10% of the time and a 20W light that is on 100% of the time. What is the energy use per month?
   a. 54Wh
   b. 68kW
   c. 68kWh
   d. 0.63 MWh

Question 1 is **c**
1HP = 746W
746W × 0.1 × 24hrs × 30 days = 53,712Wh
20W × 24hr × 30 days = 14,400Wh
53,712Wh + 14,400Wh = 68,112Wh
68,112Wh/1000 = **68kWh**

2. What kind of ground faults do most inverters detect?
   a. High voltage
   b. Medium voltage
   c. AC
   d. DC

Question 2 is **d**
Inverters detect dc ground faults

3. Current carrying ability of a conductor (wire) is called
   a. Ampacity
   b. Amps
   c. Inductance
   d. Insolation

Question 3 is **a**
Current carrying ability is ampacity

4. According to ohms law and voltage drop, the relationship between voltage and current is
   a. Proportional
   b. Inversely proportional
   c. Directly inverse
   d. Inversely direct

Question 4 is **a**

$V = IR$

If I goes up and R remains constant, then V goes up. Also with voltage drop, more current makes more voltage drop. **Voltage and current are proportional**.

5. An Array in Sacramento California in August is not producing as expected. Which of the following is NOT a likely problem?
   a. Decreased ac wire size and voltage loss
   b. String size too short
   c. String size too long
   d. Soiling

Question 5 is **c**

Late summer is hot and can be dry. **When it is hot, voltage can be low**. Too long of a string would be a problem when it too cold, not hot. Pay attention to the word NOT in a question. Also, since there are 2 answers that are opposites, there is a good chance that one of them is the correct answer.

6. When using a digital multimeter, what is the first thing you should do?
   a. Turn the meter on
   b. Inspect the meter
   c. Touch the black lead to the white wire
   d. Touch the black lead to the black wire

Question 6 is **b**

Always inspect first. When taking an exam always think on the safe side. Look before you touch.

7. Which is the worst design error?
   a. BIPV on a commercial job
   b. A fast growing tree to the north of the array in Idaho
   c. PV at 5 degree tilt at 40 degrees latitude
   d. Different tilt angles in a PV source circuit

Question 7 is **d**
Different tilt or azimuth orientation within a source circuit is bad. **Series connections should not have different orientations**.

8. The low design temperature is −20°C and the Voc temperature correction factor for your PV is −0.34%/°C. The characteristics of your module are Isc = 8A, Voc = 37V and Vmp = 29V. How many modules can you put in series for an inverter that cannot go over 500V?
   a. 10
   b. 11
   c. 12
   d. 13

Question 8 is **b**
−20°C − 25°C = −45°C
−45°C × −0.34%/°C = 15.3% increase Voc
1.15 × 37Voc = 42.6V cold
500V/42.6V = 11.7
Round down to **11 in series**
Rounding up would bring the voltage over the inverter limit.

9. MPPT is for
   a. Optimizing energy from the utility
   b. Getting the most energy from a battery with a MPPT inverter
   c. Optimizing power from a PV array
   d. Tracking the sun with a 2-axis tracker

Question 9 is **c**
MPPT is maximum power point tracking and will **optimize the energy harnessed from PV**.

10. Concentrated PV
    a. Only works with direct sunlight
    b. Usually has a single axis tracker
    c. Will get 30% power during cloudy conditions
    d. Is using heat to make electricity

Question 10 is **a**

Concentrated PV only works with direct sunlight. Also concentrated PV works with 2-axis trackers and will not work at all during cloudy conditions.

11. Fall protection should be used at heights over
    a. 4 feet
    b. 2 feet
    c. 3 feet
    d. 6 feet

Question 11 is **d**

Fall protection required over 6 feet

12. Parallel PV circuit connections increase
    a. Voltage
    b. Current
    c. Resistance
    d. EMF

Question 12 is **b**

Parallel connections increase current. Series increases voltage.

13. Optimal tilt for off-grid PV system in the winter
    a. Latitude
    b. Latitude –15 degrees
    c. Latitude +15 degrees
    d. Vertical

Question 13 is **c**

Optimal tilt to maximize winter production is latitude +15 degrees. The sun is lower in the winter, which requires a higher tilt.

14. For a transformerless inverter, which type of wire should be used when exposed to the elements outdoors on PV source circuits?
    a. PV wire
    b. THWN-2
    c. USE-2
    d. PV-2

Question 14 is **a**

PV Wire must be used for exposed **transformerless** (ungrounded) inverter source circuit wiring.

15. Bypass diodes are located in the
    a. Combiner box
    b. Inverter
    c. Wiring harness
    d. PV module

Question 15 is **d**

Bypass diodes are in PV module junction box.

16. A PV array at a higher elevation would tend to produce more
    a. Current
    b. Heat
    c. Resistance
    d. Polysilicon

Question 16 is **a**

Higher elevation = less atmosphere = more irradiance = **more current**.

17. Voc = 36V, low temp = −2°C, inverter maximum input voltage = 500V, temp coef voltage = −0.35%/°C. What is the maximum number of modules that can be connected in series?
    a. 13
    b. 10
    c. 11
    d. 12

Question 17 is **d**
−2°C − 25°C = −27°C
−27°C × −0.35%/°C = 9.45% increase Voc
1.0945 × 36Voc = 39.4Voc cold
500V/39.4V = 12.7
round down to **12 in series**

18. How much 14% efficient PV fits on 25 square meters?
    a. 3000W
    b. 4.5kW
    c. 4000W
    d. 3.5kW

Question 18 is **d**
14% efficient = 0.14 × 1000W/m² = 140 W/m²
140W/m² × 25 m² = 3500W = **3.5kW**

19. Which is the smallest conductor?
    a. 18 AWG
    b. 10 AWG
    c. 0 AWG
    d. 3/0 AWG

Question 19 is **a**
18 AWG is the smallest conductor. From small to large the conductors listed are 18AWG, 10AWG, 0 AWG, 3/0 AWG. 0 AWG is also called 1/0 AWG.

20. Which of the following is the simplest type of PV system?
    a. Utility interactive
    b. Stand-alone
    c. Self-regulating
    d. Direct coupled

Question 20 is **d**
Direct coupled is the simplest and is just PV and a load that only works when the sun is out.

21. Sun path charts differ based on
    a. Longitude
    b. Latitude
    c. Azimuth
    d. Tilt angle

Question 21 is **b**
Sun charts are specific for different latitudes (latitude = degrees from equator).

22. Series connections increase
    a. Voltage
    b. Current
    c. Cold
    d. Heat

Question 22 is **a**
Series increases voltage. Parallel increases current.

23. The longest shadow at 2PM would be on
    a. June 1
    b. November 12
    c. February 14
    d. March 29

Question 23 is **b**
The day closest to winter solstice (about 12/21) will have the longest shadow.

24. What is the ground mount advantage over a rooftop system?
    a. Takes no extra real estate
    b. Cooler
    c. Distributed generation
    d. Soiling

Question 24 is **b**
A ground mount is **cooler** than a rooftop because of better airflow and will have higher voltage.

25. Which test conditions have higher voltage?
    a. STC
    b. PTC
    c. CEC
    d. NOCT

Question 25 is **a**
STC is higher voltage. The other test conditions listed have higher cell temperatures and lower voltage.

26. Which is the best tilt angle for annual production in most of the USA
    a. Latitude +15 degrees
    b. Latitude −15 degrees
    c. 30 degrees
    d. Longitude tilt

Question 26 is **c**
Best tilt for most of the USA for annual production is **30 degrees**.

27. A 100kW STC PV system with 8% system losses, using PV that is +3 –0% production tolerance, a 95% efficient inverter, what would be the expected ac output if 1000W/m² irradiance?
    a. 84.8kW
    b. 100kW
    c. 87.4kW
    d. 88.7kW

Question 27 is **c**

100kW × 0.92 × 1 × 0.95 = **87.4kW**

For +3 –0% production tolerance the derating factor is 1, because we do not lose anything. We are not expecting to gain anything because we are careful, so we ignore the +3%.

28. A 12V light is left on for 13 hours at 1.5A with a 12V battery. How much energy is used?
    a. 23 Ah
    b. 0.234 kWh
    c. 23.4 Wh
    d. 19.5 Ah

Question 28 is **b**

12V × 1.5A × 13hr = 234Wh

234Wh = **0.234kWh**

29. In the question above, how many Ah are used?
    a. 23 Ah
    b. 0.234 kWh
    c. 23.4 Wh
    d. 19.5 Ah

Question 29 is **d**

1.5A × 13hrs = **19.5 Ah**

30. Electrical safety is best studied at
    a. OSHA website
    b. OSH website
    c. CSLB website
    d. UL website

Question 30 is **a**
Occupational Safety and Health Administration (**OSHA**) is for workplace safety including electrical safety. The National Electric Code (NEC), published by the National Fire Protection Association (NFPA), is also for electrical safety.

31. The interconnection of a PV system is approved by
    a. City
    b. County
    c. State
    d. Utility

Question 31 is **d**
Interconnection is approved by the utility.

32. Grounding (bonding) of a transformerless PV system on a roof should be done to
    a. Stainless steel hardware
    b. Flashings
    c. Racking and aluminum frames
    d. Grounded conductor

Question 32 is **c**
Racking and frames should be grounded (bonded) on every PV system, even ungrounded systems.

33. Lowest grid voltage
    a. Transmission
    b. Distribution
    c. Service equipment
    d. Generation

Question 33 is **c**
Service has the lowest voltage. Service voltage is the voltage at the building.

34. If a 12V PV module has 36 cells, then how many volts is a 54-cell module?
    a. 6V
    b. 18V
    c. 24V
    d. 17V

Question 34 is **b**
12V / 36 cells = 0.333V/cell
0.333V/cell × 54 cells = **18V**

35. Transformers convert
    a. High voltage low current to low voltage low current
    b. Low voltage low current to high voltage high current
    c. High voltage low current to low voltage high current
    d. Low current low current to high voltage low current

Question 35 is **c**
Transformers convert **high voltage low current to low voltage high current** (or vice versa).

36. Undersized PV array
    a. Decreases voltage drop
    b. Decreases battery life
    c. Increases current drop
    d. Increases resistance

Question 36 is **b**
Decreased array size decreases battery life (due to not charging battery fully). Not fully charging a battery is not good for the battery.

37. Bypass diodes
    a. Reduce current
    b. Limit effects of shading
    c. Are wired in series with groups of solar cells and prevent reverse current
    d. Are in the combiner box

Question 37 is **b**
Bypass diodes reduce effects of shading and are wired in parallel in cells in the junction box on the back of the PV module.

38. Microinverters are connected between the _____ and the _____.
    a. PV module, service panel
    b. String, combiner
    c. Main breaker, backfeed breaker
    d. Utility meter, transformer

Question 38 is **a**
Microinverters (and other **inverters**) are **between the PV and the service panel**.

39. A system that sits on a flat/low slope rooftop and has no penetrations is called a
    a. Flush mounted system
    b. Flashed system
    c. BIPV system
    d. Ballasted system

Question 39 is **d**
Ballasted systems are common flat low slope (flat) roof systems that are not penetrating the roof.

40. What is the proper slope of an extension ladder?
    a. 1:3
    b. 1:4
    c. 1:5
    d. 1:6

Question 40 is **b**
1:4 slope is required for a ladder according to OSHA.

41. Voc = 22V, Vmp = 18V, hot PV cell temp = 50°C, inverter operating voltage range = 250V to 600V, temp coefficient of voltage = −0.38%/°C. What is the least amount of modules that should be connected in series?

  a. 12
  b. 14
  c. 16
  d. 18

Question 41 is **c**

50°C − 25°C = 25°C

25°C × −0.38%/°C = 9.5% decrease in voltage

1 − 0.095 = 0.905

0.905 × 18Vmp = 16.3Vmp hot

250V / 16.3Vmp = 15.3

Round up to **16 in series**

If you round down for hot temperatures you would not have enough voltage, so round up for hot and round down for cold.

For hot voltage calculations, do not use Voc, use Vmp. Use Voc for cold calculations.

42. The primary power source for a utility interactive PV system is the
   a. Inverter
   b. Utility
   c. PV
   d. Charge controller

Question 42 is **b**

The **primary power source** for a **utility interactive** system is the **utility**. The interactive inverter will match the voltage and frequency of the utility.

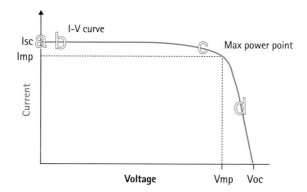

43. Which place on the IV curve above produces power and has the most current?
   a. A
   b. B
   c. C
   d. D

Question 43 is **b**

A is the most current, but produces no power and **b is the next most amount of current and produces power** (c produces the most power and d is the most voltage).

44. What is specified on a charge controller
    a. Irradiance limit
    b. Insolation limit
    c. Maximum voltage
    d. Minimum current

Question 44 is **c**

A charge controller will have **maximum voltage specified** (and so will most any equipment). A charge controller will also specify maximum current.

45. Units for irradiance are
    a. Energy per area
    b. Power per area
    c. Volts per square meter
    d. Amps per square meter

Question 45 is **b**

Irradiance is power per area, usually measured in watts per square meter. $1000W/m^2$ is the irradiance level called "peak sun". Energy per area is irradiation or insolation.

46. What is used to calculate maximum system voltage?
    a. Current
    b. Power
    c. Temperature
    d. Irradiance

Question 46 is **c**

Temperature is used to calculate maximum system voltage along with Voc and the temp coefficient of Voc.

47. What will be increased by reflective surfaces (albedo)?
    a. Voltage
    b. Resistance
    c. Current
    d. AM

Question 47 is **c**

Reflections and increased light increases current. Reflected light can also be called **albedo**.

For the following 3 questions, use the information from the table below:

Question 48 to 50 Table

Power	250W
Open Circuit Voltage	40V
Maximum Power Voltage	35V
Short Circuit Current	8A
Maximum Power Current	7.14A
Temp Coefficient for Voltage	−0.33%/°C
CEC Power	220W
NOCT	48°C
Inverter low MPPT voltage	230V
Inverter maximum input voltage	550V

48. There are 4 PV source circuits of 10 modules each. What is the MPPT voltage of the array at Standard Test Conditions?

    a. 160V

    b. 350V

    c. 140V

    d. 230V

Question 48 is **b**

35Vmp × 10 in series = 350Vmp

49. What is the power of the array?

    a. 10kW

    b. 110kW

    c. 1000W

    d. 40kW

Question 49 is **a**

250W × 40 modules = 10,000W

10,000W = **10kW**

An easy trick is to convert the power of the PV module to kW immediately, so a 250W module would be 0.25kW; usually we just put the decimal in front.

50. What is the maximum number of modules in series at a location with a low temperature of minus 40°C?
    a. 9
    b. 10
    c. 11
    d. 12

Question 50 is **c**
−40°C − 25°C = −65°C
−65°C × −0.33%/°C = 21.45% increase in Voc
1.2145 × 40V = 48.6V
550V / 48.4V = 11.4
11 in series max

51. What is the best voltage for charging a 12V battery?
    a. 12V
    b. 11.8V
    c. 12.3V
    d. 14.1V

Question 51 is **d**
Best voltage for **charging a 12V battery is 14.1V** (best of the choices given).

52. Unit for resistance?
    a. Amp
    b. Volt
    c. Om
    d. Ohm

Question 52 is **d**
Resistance is measured in Ohms (Ω).

53. Wind uplift forces in the construction industry in the US are measured in
    a. Pounds per square foot
    b. Miles per hour
    c. Square feet per pound
    d. kW per m^2

Question 53 is **a**
Wind uplift is measured in pounds per square foot (psf).

54. Direction on a compass is
    a. Latitude
    b. Longitude
    c. Meridian
    d. Azimuth

Question 54 is **d**
Direction on a compass is azimuth.

55. Hydraulic analogy for voltage is
    a. Pressure
    b. Flow
    c. Volume
    d. Drag

Question 55 is **a**
Hydraulic analogy for voltage is pressure. Current is flow.

56. Which room cannot have a PV dc disconnect?
    a. Kitchen
    b. Attic
    c. Bathroom
    d. Bedroom

Question 56 is **c**
PV dc disconnect not allowed in bathroom (wet steam).

57. Stand-alone inverter output goes to
    a. Loads
    b. Utility
    c. Meter
    d. PV

Question 57 is **a**
Off-grid inverter output goes to loads.

58. Best sealants for residential rooftop
    a. Polyurethane and silicone
    b. Tar and paraffin
    c. Bitumen and rubber
    d. Silicon and caulk

Question 58 is **a**
Polyurethane and silicone are the best sealants of the examples given.

59. 500W/m^2 for 4 hours is
    a. 20kWh
    b. 2kWh
    c. 2kW
    d. 5kW

Question 59 is **b**
500W/m^2 × 4 hours = **2kWh**

60. For a supply side connection, the output of the inverter is limited by
    a. Service entrance conductors
    b. Load side breaker
    c. PV dimensions
    d. Combiner box ampacity

Question 60 is **a**
Supply side inverter output is **limited by the service entrance conductors**. The loads are already protected by the main service disconnect (main breaker).

61. Which is the smallest installable part of a PV system?
    a. Solar panel
    b. Solar module
    c. Solar array
    d. PV source circuit

Question 61 is **b**.
A solar module is the smallest unit that you can install. A solar panel is made up of solar modules.

62. What is the typical hazard with PV systems?
    a. Shock
    b. Arc flash
    c. Lightning
    d. Heat stroke

Question 62 is **a**.
Solar installers are always concerned about shocks and falls.

63. A 4kW PV system with 15% losses in a location with 4.5 sun hours per day, what would be the annual output?
    a. 985.5kWh
    b. 1450kWh
    c. 5585kWh
    d. 4550kWh

Question 63 is **c**.
4kW × 0.85 × 4.5 sun hours × 365 days = 5585kWh
Losing 15% means keeping 85%, which is where the 0.85 derating factor comes from. Always think about what you keep.

64. If available roof area is 12m^2 and the PV is 15% efficient, then how much PV will fit?
    a. 4.5kW
    b. 2.2kW
    c. 1.8kW
    d. 3.3kW

Question 64 is **c**.
15% efficient PV is 15% of 1000W/m^2
$0.15 \times 1000W/m^2 = 150W/m^2$
$150W/m^2 \times 12m^2 = 1800W$
1800W = **1.8kW**

65. Which of the following is a typical direct PV system application?
    a. Lighting
    b. Batteries
    c. Pumping water
    d. Fire alarm

Question 65 is **c**.
Pumping water is a common direct-coupled PV application. A water pump motor hooked up to PV with no inverter or battery is a direct system. When the sun shines, the pump will pump.

66. Current is affected mostly by
    a. The number of solar cells in series
    b. The size of the solar cell
    c. The backsheet of the solar module
    d. The temperature of the cell

Question 66 is **b**.
The size of the solar cell determines the current of a PV module. The number of cells in series determines the voltage.

67. In California the north needle of the compass will point
    a. East of true north
    b. West of true north
    c. North of true north
    d. South of true north

Question 67 is **a**.
In California magnetic declination is to the east.

68. Maintenance of sealed valve regulated lead acid batteries includes
    a. Checking fluid levels
    b. Equalization charge
    c. Scrubbing lead plates
    d. Cleaning terminals

Question 68 is **d**.
On a sealed battery, one does not have access to the electrolyte fluids.
Equalization is also not done on sealed batteries. All batteries can have their
terminals cleaned.

69. Given that Voc is 18V and the temperature correction factor = 1.2. What
    would be the maximum system voltage if there were 16 modules in series?
    a. 340V
    b. 346V
    c. 240V
    d. 440V

Question 69 is **b**.
18V × 1.2 correction factor × 16 in series = 346V

70. STC watts can be calculated by
    a. Vmp × Voc
    b. Voc × Isc
    c. Vmp × Imp
    d. PTC × CEC

Question 70 is **c**.
Vmp × Imp = STC watts
At Voc there is no current and power is zero.
At Isc there is no voltage and power is zero.

CHAPTER 3 PV MATH ANSWERS

1. A 200-watt light is left on for 7 days. How much energy is consumed?
   a. 34kWh
   b. 336kW
   c. 3360Wh
   d. 14MWh

power × time = energy
200W × 24 hrs/day × 7 days = 33,600Wh
33,600Wh/1000W/kW = 33.6kWh
33.6kWh is closest to 34kWh
correct answer is a.

2. If a 100W light bulb is working on a 120V socket, what is the current?
   a. 12kW
   b. 0.83A
   c. 1.2A
   d. 144 Ω

volts × current = power (VI = P)
I = P/V
I = 100W/120V = 0.833A
correct answer is b.

3. How much resistance does the light bulb have in question 2 above?
   a. 0.83 ohms
   b. 12 ohms
   c. 0.007 Ω
   d. 144 Ω

volts = current × resistance (V = IR)
R = V/I
R = 120V/0.833A = 144Ω
correct answer is d.

4. The hydraulic analogy for voltage is
   a. Flow
   b. Volume
   c. Capacity
   d. Pressure

voltage is like pressure
correct answer is d.

5. The hydraulic analogy for current is
   a. Flow
   b. Volume
   c. Capacity
   d. Pressure

current is like flow
correct answer is a.

6. The hydraulic analogy for energy is
   a. Flow
   b. Volume
   c. Laminar
   d. Pressure

energy is like volume
correct answer is b.

7. 2000 watts is equal to
   a. 20kW
   b. 2kW
   c. 0.2kW
   d. 2kWh

kilo = 1000
1kW is 1000W, so 2kW is 2000W (move decimal 3 places).
correct answer is b.

8. The power of a 250-watt solar module is equal to
   a. 1/4th of a kW
   b. 0.025MW
   c. 25KVA
   d. 0.25kWh

move decimal 3 places to left to make watts kilowatts
250W = 0.250kW
0.25 =25%= ¼
correct answer is a.

9. One horsepower equals
   a. 0.746kW
   b. 746kWh
   c. 1000W
   d. 1MW

1 HP = 746W
746W/ 1000W/kW = 0.746kW (move decimal 3 places)
correct answer is a.

10. A pump works at 4A and 12V for 3 hours. How much energy does it consume?
    a. 144kWh
    b. 48Wh
    c. 0.144kWh
    d. 96Wh

Volts × Current = Power (VI = P)
12V × 4A = 48W = Power
Power × Time = Energy
48W × 3 hours = 144Wh
144Wh/ 1000W/kW = 0.144kWh
correct answer is c.

When taking an exam, the best way to perform well is by being prepared, calm, confident, and well-rested. Good luck!
Sean White